Max Parsonage

CHEMISTRY

FACTS & PRACTICE FOR A LEVEL

OXFORD
UNIVERSITY PRESS

OXFORD
UNIVERSITY PRESS

Great Clarendon Street, Oxford OX2 6DP

Oxford University Press is a department of the University of Oxford.
It furthers the University's objective of excellence in research, scholarship, and education by publishing worldwide in

Oxford New York

Athens Auckland Bangkok Bogotá Buenos Aires Cape Town Chennai Dar es Salaam Delhi Florence Hong Kong Istanbul Karachi Kolkata Kuala Lumpur Madrid Melbourne Mexico City Mumbai Nairobi Paris São Paulo Shanghai Singapore Taipei Tokyo Toronto Warsaw

with associated companies in Berlin Ibadan

Oxford is a registered trade mark of Oxford University Press in the UK and in certain other countries

First published 2001

British Library Cataloguing in Publication Data

Data available

ISBN 0 19 914764 7

Typeset by Magnet Harlequin

Printed in England

Author's Acknowledgements

I am pleased to recognise the special contributions of a few, amongst many others, that have made this book more accurate, clear, and accessible. My thanks go to my wife Jane for her checking and encouragement, Dr. Steve Field for checking the chemical accuracy and question answers, and Peter Mellett for ensuring clear text and precise chemistry.

I am grateful for the comments and enthusiasm of many students, including Toby and Hannah.

INTRODUCTION

My aim in writing this book was to make chemistry both concise and clear, so that students can quickly reinforce class work, and subsequently revise it. Through my role as an A Level Chemistry tutor I have been able to test successive drafts of this material with students, and this has been extremely valuable and interesting. It was a huge challenge to condense the essence of the subject into so few pages, but I know from working with students the value of doing this. There is just as much emphasis in the book on questions as there is on content, as only by testing themselves do students really discover how good their understanding of a topic is.

Exam-style questions are provided for students to practise their exam technique. A variety of styles is presented to reflect the diversity of questions in the papers. Students are given realistic space to write their answers, as in real exam questions. Students and teachers will be able to diagnose any exam technique problems by looking over the written answers.

MAX PARSONAGE

Max Parsonage is Head of Chemistry at d'Overbroeck's, an independent school in Oxford. He also writes about science for children, and develops interactive educational science software.

CONTENTS

WHY STUDY CHEMISTRY?

Chemistry is fundamental to understanding the world around us, simply because everything is made of chemicals. From planets to cosmetics, and microbes to bridges, chemistry underpins how materials behave. It also explains how different substances can be made.

If you study AS or A level chemistry then you should be able to ask 'Why?' and receive satisfying explanations. You will find AS chemistry explains chemical ideas mostly using words, while A2 chemistry explains chemical ideas using maths as well as words.

If you like logic problems, the way ideas can just 'click' beautifully together, then you will enjoy chemistry. Once you have gained a good grasp of the chemical patterns you will find there is very little detail to memorise, because studying chemistry is like studying a game. Once you know the 'rules of chemistry', you can 'play' with the chemical ideas. Chemistry is therefore a concise subject. It is attractive because it makes you think, without requiring you to write many essays or memorise huge amounts of information. Studying chemistry complements A levels that are essay based or require a huge reading load.

You may have to take chemistry if you want to become a doctor, or vet, or if you want to study chemistry, pharmacology, environmental science, or related subjects. Chemistry is also a useful foundation for biochemistry, geology, physical geography, engineering, or materials science at university.

Because it is such a fundamental study, chemistry provides helpful background for a great variety of subjects, such as biology, pyrotechnics in theatre studies, and art restoration. If law interests you, then chemistry is a useful discipline because it encourages logical thinking.

HOW TO USE THIS BOOK

This book will supplement a standard A level textbook, or you could use it as a free-standing A level book to consult alongside your notes. It is best used as a course companion, to be referred to when starting a topic, and to help you understand when you are in difficulties. When tests and exams approach it will usefully explain things in a few pages, and test you thoroughly.

Teachers may set the book as homework, or use it in class tests.

HOW TO SUCCEED IN YOUR STUDIES

To succeed in A level chemistry you, the student, need to **understand** the ideas, **remember** the facts and ideas, and have a good **exam technique**.

An understanding of chemistry builds up layer upon layer, so the units are laid out with the foundation topics first. The units are presented in the best order for you to study them. Unit 1 (Bonds and structures) will help you understand many other topics. Similarly the early topics will help with the later ones when studying Energetics, Equilibrium, Kinetics and Organic chemistry. Understanding groups 1, 2, and 7 will help you with transition metals.

Cover one topic at a time to gain a full understanding, rather than scanning many topics quickly.

To **understand** the topics, read the relevant section. If you do not immediately understand an explanation, pause and re-read it. Use the examples to help you see the point. By writing short answers to the recall questions, test yourself to check you understand the ideas, and then refer to the answers.

To **remember** the facts and ideas, use the factual recall questions (the 'Recall test'). If you can answer all the questions in a section, then you have the facts you need to answer the exam questions on that topic (the 'Concept test'). These questions are a more effective way of memorising information than simply copying notes. You could use them to help you to identify your weaknesses, then return to the unit itself to turn your weaker topics into strengths.

HOW TO SUCCEED IN EXAMS

When you understand the concepts and have memorised any necessary ideas, you can work on improving your **exam technique**.

To gain marks in exams you must, of course, understand and know the topics. You must also have an effective exam technique; remember to read the questions carefully, make sure you understand what the examiners are asking, answer the question (rather than just writing anything you know), and communicate a clear answer using technical words correctly.

Students commonly lose marks by not reading the questions. It may appear obvious, but in the stress of the exam many students do not always read every word and so do not answer a question appropriately. So always read the question at least twice. Many students write everything they know about the subject mentioned in the question, and so produce long-winded answers. This may waste so much time that they do not finish the exam paper. More importantly, examiners state that long rambling answers tend not to gain full marks because they do not focus on the particular point raised in the question.

The examiners use particular 'command words' which indicate how you should respond to questions; marks are easily lost by ignoring these. For example, many students 'explain' when the question says 'describe'. Here is a list of command words found in A-level chemistry exam papers.

Define or *What is meant by...* – Give a definition in words and in equations if possible.

Describe – This asks you to state what is observed in an experiment, or state the basic points in a practical method. Giving a chemical explanation is not necessary.

Describe what will be observed – 'Observed' means seen, or sensed, so describe colours, states, and smells. Chemical names may not gain you marks.

Explain – say how and why something happens. Be careful to use the correct technical words. If many marks are offered then explain in depth.

Calculate – Obviously, work out a numerical answer. Remember to give the correct sign (for example, exothermic enthalpies are negative), and give your answer to the correct number of significant figures.

Using the data given – You must refer to the data given! Show the examiner you have done so by marking graphs, using figures in calculations, or using words from the question. Be wary of basing your answer on recall of knowledge.

Give the formula – You must give the formula, not the name. Easily overlooked.

Name – Give the name. The examiner is checking that you can name compounds. A formula will *not* do.

Identify – Give the chemical name or formula.

Suggest – Anything reasonable will do as an answer. There are many possible responses. This confuses some students because it is so unusual in chemistry exams; the examiner usually wants a particular answer.

Comment on – The examiner wants you to point out an idea, usually in the specification (syllabus), suggested by the data.

Here are common technical words that students misuse. Take care not to confuse them. Examiners will not give marks if you talk about atoms in sodium chloride – because it contains ions, of course.

Words for particles are *atom*, *ion*, and *molecule*.

Chemical substances are *elements*, *compounds*, or *mixtures*.

Bonding must be *covalent*, *ionic*, or *metallic*...

...Whereas structures must be *simple covalent molecules* or *giant lattices* (which could be *covalent ionic*, or *metallic*).

THE A LEVEL SYSTEM

The system is designed so that a typical student will study four or five ASs in their first year of study, and then select three from these to continue in their second year of study, called A2. This will then give them three full A levels. The system is designed to be flexible, so that it is possible to do any number of ASs and A2s, but an A2 can only be done if the relevant AS has been completed. In reality, most students' choices will be limited by whatever system and options their school or college can offer.

Two of the three main examination boards are offering two specifications (syllabuses) in chemistry, and one main board is offering one specification. Each AS component contains three modules, as does each A2, but the content of each module varies from specification to specification. Whatever specification a student studies, most of the content covered is the same as in any other specification, but the topics are mixed differently in the different modules.

EXAM BOARD SPECIFICATIONS AND MODULES

Module	1 AS	2 AS	3 AS	4 A2	5 A2	6 A2
Exam Board						
AQA	1, 2, 3, 5, 29	4, 11, 13, 15, 16	7, 8, 9, 10, 28	8, 14, 17, 18, 19, 20, 21, 22, 23	6, 11, 12, 22, 23, 24, 25, 26	27, 28
EDEXCEL	1, 2, 3, 4, 5, 29	7, 8, 9, 10, 11, 13, 15, 16	3, 4, 28, 29	6, 8, 12, 17, 18, 20, 21, 22	8, 14, 19, 23, 24, 25, 26, 28	27, 28
EDEXCEL (Nuffield)	1, 2, 3, 8, 10, 11, 29	1, 2, 4, 8, 9, 10, 11, 15, 16	28	12, 13, 14, 17, 18, 19, 20, 21, 23	27, 28	12, 22, 23, 24, 25, 27
OCR A	1, 2, 3, 4, 5, 29	7, 8, 9, 10	11, 13, 15, 16, 28	7, 8, 19, 20, 21, 22, 23	6, 7, 12, 22, 24, 25, 26, 27	14, 17, 18, 27, 28
OCR (Salters)	1, 5, 6, 8, 9, 11, 12, 29	1–5, 7, 8, 9, 10, 15, 16, 22, 23	28	7, 8, 14, 17, 21, 22, 23, 24, 25, 26	6, 8, 12, 15, 17, 18, 19, 20, 23, 27	27, 28
WJEC	1, 2, 3, 4, 5, 29	7, 8, 9, 10, 11, 13, 15–18	28	7, 8, 19, 20, 21, 22, 23	3, 4, 6, 12, 14, 18, 24, 25, 26	27, 28
CCEA	1, 2, 4, 5, 11, 28, 29	3, 7–10, 13, 15, 16, 28, 29	28	6, 12, 14, 17, 18, 20, 21, 22, 26, 29	19, 20, 21, 23, 24, 25, 29	27, 28

The relevant units in this book are shown for each module of each specification.

Unit 1

BONDS AND STRUCTURES

Every substance is made of atoms. The arrangement of atoms and the bonding between them determines the physical and chemical character of all substances.

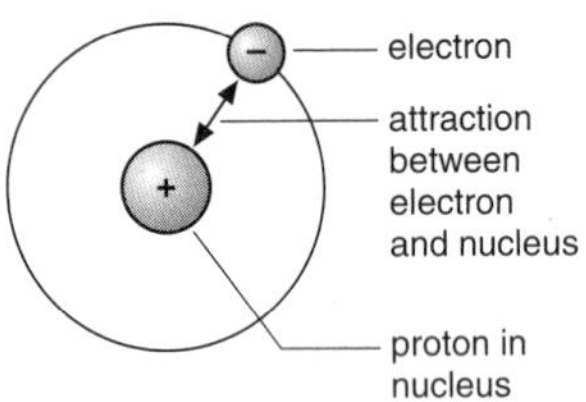

Fig. 1.1

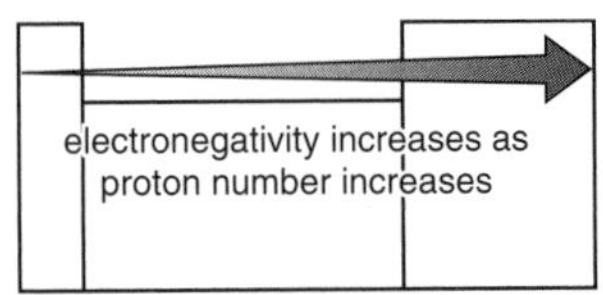

Fig. 1.2

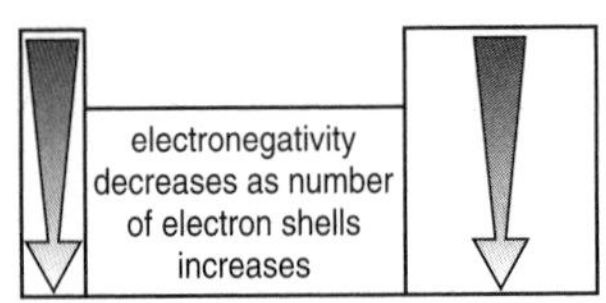

Fig. 1.3

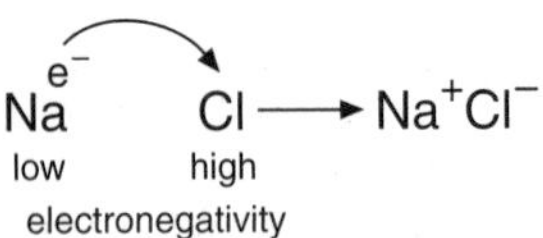

Fig. 1.4

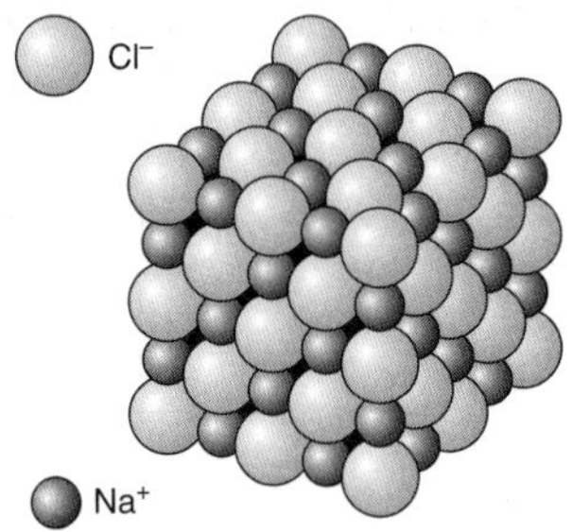

Fig. 1.5

Fig. 1.6

- Atoms have structure. The central **nucleus** contains positively charged protons (+) and neutral neutrons (0) and is surrounded by layers, called **shells**, of negatively charged electrons (–). The electron number is equal to the proton number (atomic number) in a neutral (uncharged) atom. All the atoms in a given element have the same proton number. Atoms of different elements react differently, according to the actual electron number in their shells. See Fig. 1.1.

- Elements are arranged in the **periodic table** in order of increasing proton number. The proton number increases from left to right across horizontal rows (called **periods**) and from the top to the bottom of vertical columns (called **groups**). Look at the periodic table printed in this book as you study the next section.

  Across a period, increasing proton number increases the attraction of the nucleus for the electrons in the outer shell. Therefore, **atomic radius** decreases as proton number increases across a period.

  Down a group, the number of electron shells increases as proton number increases. Filled inner shells increase the distance of the outer shell from the nucleus. These filled shells also **shield** electrons in the outer shell from the attraction of the full nuclear charge. Therefore, atomic radius increases as proton number increases down a group.

- **Electronegativity** is a measure of how attractive an atom is for a pair of electrons in a covalent bond. An element's attracting ability increases with increasing electronegativity value. See Fig. 1.2 and Fig. 1.3.

  Electronegativity values affect the type of bond that forms.

- **Ionic bonds** occur when one atom is much more electronegative than another atom. The atom with the smaller electronegativity loses an electron to the atom with the greater electronegativity. The electron acceptor gains (–) electrons and so becomes a negative ion (**anion**) while the electron donor loses electrons to become a positive ion (**cation**). See Fig. 1.4.

  Anions and cations combine in a regular pattern to form **giant ionic lattices**. Melting points are high due to the strong electrostatic forces between the oppositely charged ions. See Fig. 1.5 and Fig. 1.6. **Ionic bond strength** is greatest when the ions have large charges and small ionic radii.

  Ionic compounds are **brittle**. External forces displace ions so that similar charges are next to each other. They repel and the lattice breaks.

- **Metals** have relatively small electronegativity values. **Metallic bonds** form when weakly held electrons in the outer shell become mobile and move freely between atoms. The resulting metal cations are bonded together by their attraction for the mobile (**delocalised**) electrons. They arrange in regular patterns to form **giant metallic lattices**. See Fig. 1.7.

  Metallic bond strength, and hence hardness and melting point, generally increase with increasing numbers of delocalised electrons. Smaller atoms make for stronger structures.

- **Non-metals** have relatively large electronegativity values. Atoms form **covalent bonds** by sharing electrons (see Fig. 1.8). Covalent bond strength generally increases with increasing electronegativity. **Dative covalent (co-ordinate) bonds** occur when both electrons in a covalent bond come from only one of the atoms.

- There are two sorts of covalent structures. **Giant covalent** (**macromolecular**) **lattices** (e.g. diamond) tend to be very hard because each atom has a strong covalent bond to the next atom in the structure. **Simple covalent molecules** (e.g. water and chlorine) have strong covalent bonds between the atoms that make up each molecule, but there are only weak forces of attraction between the molecules. See Fig. 1.9.

- Ionic bonding and covalent bonding represent two extreme forms of **bonding character**. Ionic compounds have a degree of covalent character due to incomplete transfer of electrons. Covalent compounds have a degree of ionic character due to unequal sharing of electrons.

  All the giant lattices (metallic, ionic, and covalent) have high **melting points** and **boiling points** because large amounts of energy are needed to break the bonds and separate the atoms or ions.

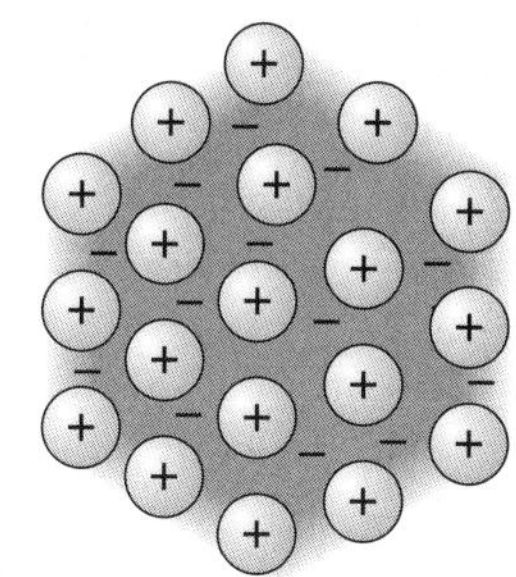

Fig. 1.7

- There are three types of **intermolecular forces** of attraction between simple covalent molecules; Van der Waals forces, permanent dipole interactions, and hydrogen bonding. Molecules with hydrogen bonding are more attractive than those with just a permanent dipole, which are more attractive than those with just Van der Waals forces.

  Weak **Van der Waals forces** (which you may know as fluctuating dipole, or London forces) exist between all molecules. You do not have to explain the origin of this force. These forces (and so the melting and boiling points) increase with increasing numbers of electrons and molecular length, e.g. methane CH_4 (M_r = 16) b.p. (boiling point) = –161 °C; straight-chain butane C_4H_{10} (M_r = 58) b.p. = –0.5 °C; globular methylpropane C_4H_{10} (M_r = 58) b.p. = –11.7 °C. See Fig. 1.10. (M_r is the **relative formula mass**, calculated by adding up the **relative atomic masses** of the elements in the formula.)

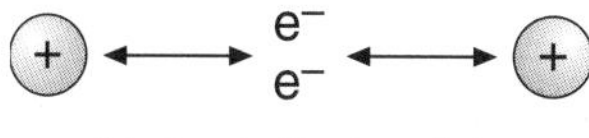

Fig. 1.8 A covalent bond

  Permanent dipole interactions exist between molecules that have **polar covalent bonds**. This type of intermolecular force results when atoms in a covalent bond have different electronegativities so that bonding electrons are shared unequally between them. **Example:** In H-Cl the Cl atom is much more electronegative than the H atom so the covalent bond is polar $^{\delta+}$H-Cl$^{\delta-}$ (see Fig. 1.11). NB In CCl_4, all the C-Cl bonds are polar, but the molecule is symmetrical, making the overall molecule non-polar. You will mainly meet permanent dipoles in covalent molecules that contain halogens or C=O or C-O bonds.

weak force

Cl — Cl ◄·········► Cl — Cl

strong bonds

Fig. 1.9

  If you compare different molecules of similar M_r, those with polar bonds (and hence permanent dipole interactions) have higher melting points than those with Van der Waals forces only.

> **Relative molecular mass** (RMM) is sometimes used instead of M_r.

  Hydrogen bonds form when two conditions are satisfied: (1) a hydrogen atom is covalently bonded to a highly electronegative atom, so that it becomes electron deficient; (2) a small strongly electronegative atom with a **lone pair** (see left) is present e.g. N, O, or F atoms. (They spell NOF). Hydrogen bonds form between the negative charge associated with the lone pair of electrons and the δ+ H atom. **Example:** Ethanol (see Fig. 1.12) and water have H-O bonds so the H atom is electron deficient; the O atom has lone pairs of electrons. By contrast, all the H atoms in ethanal CH_3CHO are joined to C atoms so none is electron deficient. Hydrogen bonding is not present in ethanal, although there are permanent dipole interactions between $^{\delta+}$C=O$^{\delta-}$ bonds. So ethanol has a higher boiling point than ethanal.

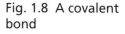

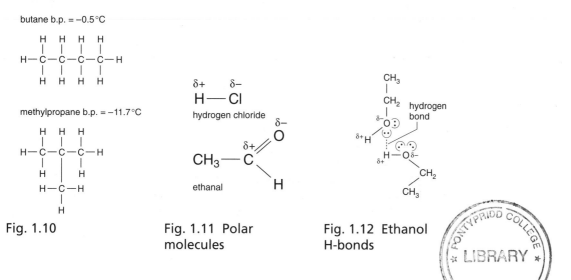

> A **lone pair** is a non-bonding electron pair in an atom's outer shell, drawn like this: (··)

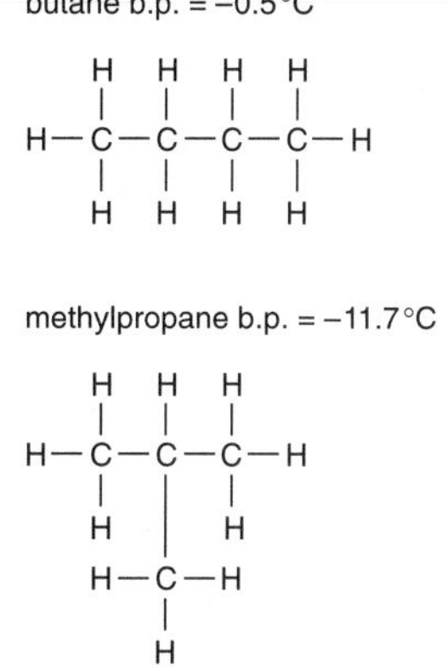

Fig. 1.10

δ+ δ−
H—Cl
hydrogen chloride

δ−
O
δ+
CH_3—C
H
ethanal

Fig. 1.11 Polar molecules

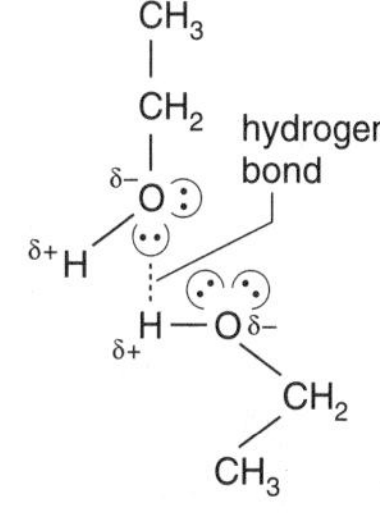

Fig. 1.12 Ethanol H-bonds

Unit 1

TESTS

RECALL TEST

1 What determines the chemical properties of an element?

(1)

2 State and explain the trend in electronegativity across the third row of the periodic table.

(2)

3 State and explain the trend in electronegativity down the second group of the periodic table.

(3)

4 Why is AgI covalent?

(1)

5 Why are some bonds polar?

(1)

6 What are the three intermolecular forces?

(1)

7 If the covalent bonds are so strong why is chlorine a gas?

(1)

8 If the covalent bonds in I_2 are weak why is iodine a solid?

(1)

9 Why does chloromethane have a dipole?

(1)

10 What is necessary for hydrogen bonding to occur?

(4)

11 Why does ethanol have a higher boiling point than ethanal?

(2)

12 Why does NaCl conduct when molten, but not when solid?

(2)

13 Why do the boiling points of the noble gases increase with increasing atomic number?

(1)

14 Why does water have a higher boiling point than the rest of the group 6 hydrides?

(1)

15 When chlorine gas is cooled, why does it condense into a liquid?

(4)

16 When solid NaCl is heated from room temperature it melts. Explain the change.

(4)

(Total 30 marks)

CONCEPT TEST

1 a Define the term 'electronegativity'.

(2)

b State and explain the bonding in carbon dioxide.

(2)

c Glucose molecules have many -OH groups. Why is glucose soluble in water?

(2)

d One explanation for the covalent character of beryllium chloride is that the covalent bonds are due to the Be^{2+} cation polarising the Cl^- anion so that the electrons in the anion are shared by the cation).

i Why does Be^{2+} polarise Cl^- when Ca^{2+} cannot?

(1)

ii Why is aluminium chloride covalent?

(2)

iii State and explain the difference in size between a chloride ion (Cl^-) and a fluoride ion (F^-).

(2)

iv Why is aluminium fluoride ionic?

(1)

2 For many years people have known two types of carbon: diamond and graphite. A third form was discovered recently containing C_{60} molecules, called buckminsterfullerene, or 'bucky balls' for short. It contains carbon atoms in a ball, a bit like a modern football, made of hexagons and pentagons. In the molecule of C_{60} the carbon atoms would be where three of the shapes join.

a Explain why graphite is soft.

(2)

b Explain why diamond is strong.

(2)

c Knowing C_{60} is made of balls, predict the properties of C_{60} at room temperature and pressure.

(2)

d One idea is to trap K^+ ions inside the C_{60} to make KC_{60}^+ ions. These cations could be combined with Cl^- ions. State and explain the magnitude of boiling point that this substance would have.

(2)

(Total 20 marks)

Unit 2 MASS SPECTROMETER AND SHAPE

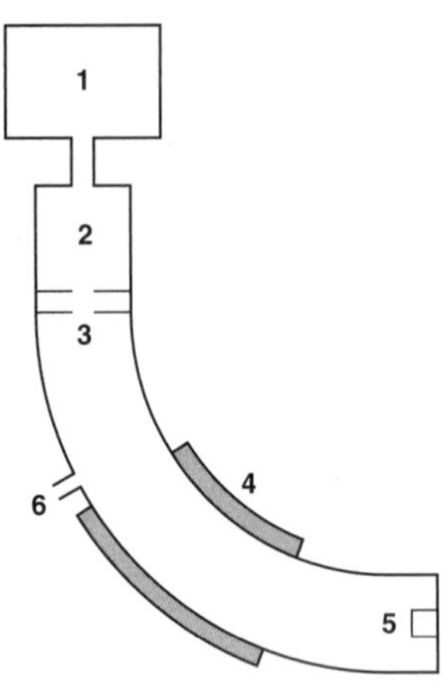

Fig. 2.1 A mass spectrometer

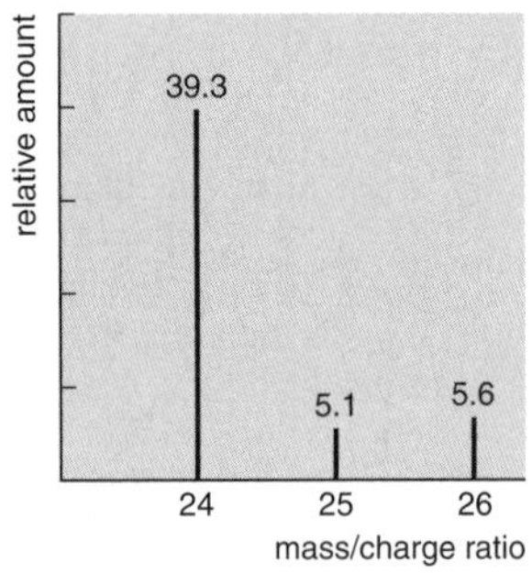

Calculation:

24 × 39.3 = 943.2
25 × 5.1 = 127.5
26 × 5.6 = 145.6
50 1216.3

1216.3 ÷ 50 = 24.326

relative atomic mass (RAM) = 24.3

Fig. 2.2 Graph of the mass spectrum of magnesium

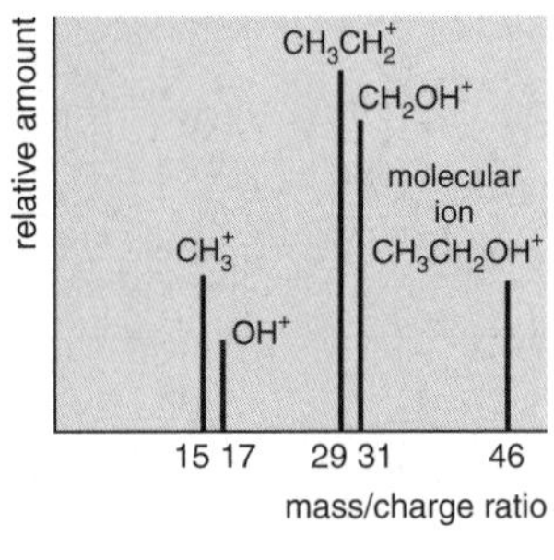

Fig. 2.3 Mass spectrum of ethanol (CH_3CH_2OH)

- There are six main points for you to remember about the functioning of a mass spectrometer (see Fig. 2.1).
  1. The sample is **vaporised** – to make the molecules or atoms mobile.
  2. High-energy electrons bombard the molecules or atoms in the vapour. Electrons are ejected and **cations form**.
  3. The cations are **accelerated** and **focused** into a beam by **electric fields**.
  4. The cation beam is **deflected** by a **magnetic field**. The angle of deflection increases with decreasing cation mass. The magnetic field strength is varied so that ions of known mass-to-charge ratio enter the **detector**.
  5. Ions entering the detector cause a **current** to flow in an external circuit connected to a **recording device**.
  6. A pump maintains a **vacuum** inside the apparatus. Otherwise the cations would be scattered by air molecules.

- You must remember these definitions:

  Atomic number (*Z*): The number of protons in the nucleus of an atom.

  Mass number (*A*): The total number of protons and neutrons in the nucleus.

  Isotopes: Atoms of the same element with different mass numbers (i.e. isotopes have the same atomic numbers but different mass numbers).

  Relative atomic mass: The weighted average mass of the atoms in a sample of an element divided by $\frac{1}{12}$th of the mass of an atom of the carbon-12 (^{12}C) nuclide.

  Relative isotopic mass: The mass of the atoms in a sample of an isotope divided by $\frac{1}{12}$th of the mass of an atom of the carbon-12 (^{12}C) nuclide.

  You must be able to calculate relative atomic mass from **isotopic mass** and **relative abundance** data (usually obtained from a mass spectrometer).

  Lay out your calculation the same way each time and then you will make fewer mistakes. **Example:**

Isotope	Abundance
^{63}Cu	69.1%
^{65}Cu	30.9%

Mass of 100 atoms = (63 × 69.1) + (65 × 30.9) = 6361.8

$$\text{Average mass of 1 atom} = \frac{\text{total mass of the atoms}}{\text{the number of those atoms}} = \frac{6361.8}{100}$$

$$= 63.6 \text{ (to 3 significant figures)}$$

You must add up the abundances to find the total number of atoms, not the masses. The total abundance may not be 100, especially if you have to measure the value from a graph (called a '**mass spectrum**'). **Example:** See Fig. 2.2.

You may see a mass spectrum of a compound that consists of **molecules**. Often, some of the original molecules are detected as well as fragments of that molecule. If the original molecule is detected, it is called the '**molecular ion**' or '**parent ion**'. The molecular ion will have the largest mass so, usually, the greatest mass (not the greatest abundance) on the spectrum will indicate the relative molecular mass of the molecule. **Example:** See Fig. 2.3.

- **Molecular shape** can be decided by using **VSEPR** (**valence shell electron pair repulsion**) theory, which depends on the number of **bonding** pairs and **lone** (non-bonding) pairs of electrons surrounding the central atom.

  Firstly, identify bonding and lone electron pairs by drawing a **dot and cross** diagram showing the **outer** (**valence shell**) **electrons** of each atom. Remember that negative ions have one or more extra electrons and that positive ions lack one or more electrons.

Molecular shape results from the electron pairs **repelling** each other so that they are as **far apart** as possible. You should expect to explain this every time an exam question discusses shape.

Lone pairs repel more than bonding pairs, so lone pairs tend to push the bonding pairs together. Repulsion decreases in the order: lone pair–lone pair > lone pair–bond pair > bond pair–bond pair.

Double bonds behave as single bonds, but with increased electron density and therefore an increased repulsive effect. For examples of molecular shapes, see Fig. 2.4.

Formula	Pairs of electrons	Bonding pairs	Lone pairs	Dot and cross diagram	Drawn shape	Name of shape	Similarly shaped ions
$BeCl_2$	2	2	0	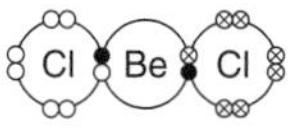	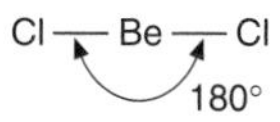 180°	linear	
BF_3	3	3	0	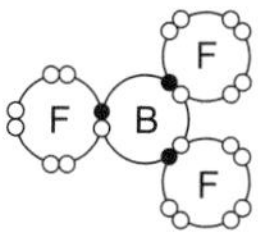	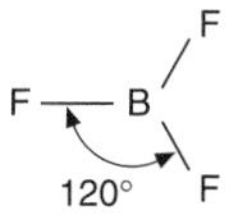 120°	trigonal planar	NO_3^- CO_3^{2-}
CH_4	4	4	0	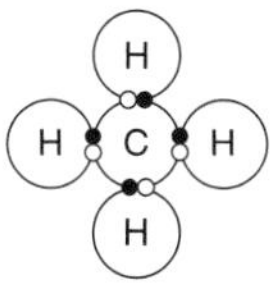	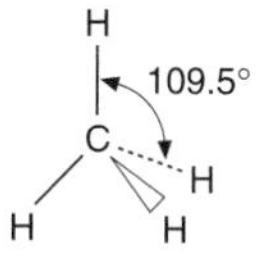 109.5°	tetrahedral	PCl_4^+ SO_4^{2-} NH_4^+
NH_3	4	3	1	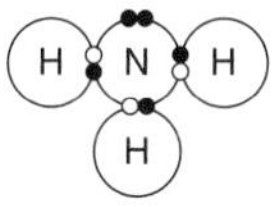	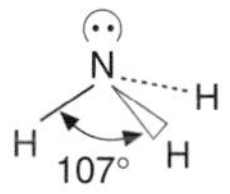 107°	pyramidal	SO_3^{2-}
H_2O	4	2	2	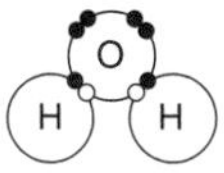	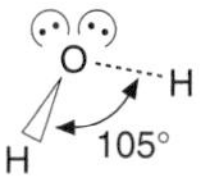 105°	V-shaped	
PCl_5	5	5	0	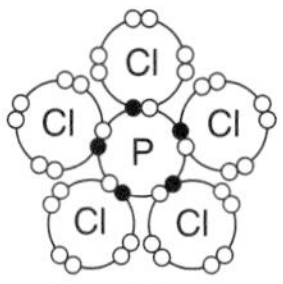	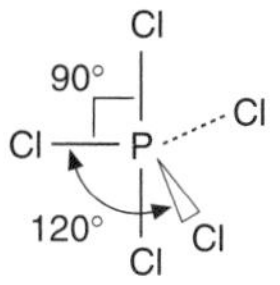 90° 120°	trigonal bipyramidal	
SF_6	6	6	0	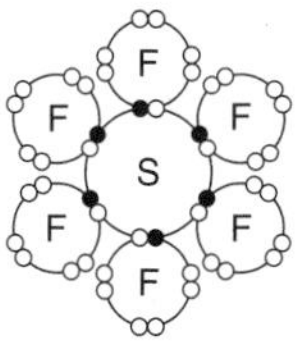	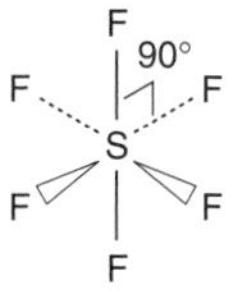 90°	octahedral	PCl_6^-

Fig. 2.4

shape of ethene ~120°

shape of SO_2

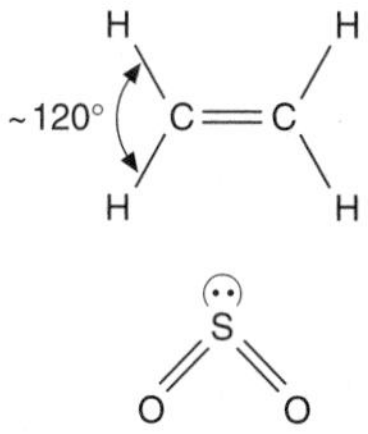

Fig. 2.5

Carbon dioxide CO_2 is **linear**; ethene CH_2CH_2 is **trigonal planar** around each carbon atom; SO_2 is **V-shaped** because of the presence of a lone pair (see Fig. 2.5).

- Always carry out the steps shown in the following example:

  Question: What is the shape of XeF_4?

  (i) The number of electrons in the outer shell of the Xe atom is eight; there are seven outer-shell electrons in each F atom.

  (ii) Each F atom shares one electron with the Xe atom to make one covalent bond.

  (iii) Four Xe electrons are used to bond with the four F atoms.

  (iv) Four of the Xe electrons are not used for bonding and exist as two lone pairs.

  (v) There are six electron pairs around the Xe atom, so the shape of XeF_4 is based on an octahedron. Lone pairs are opposite each other, not adjacent (see Fig. 2.6) so the actual shape is square planar.

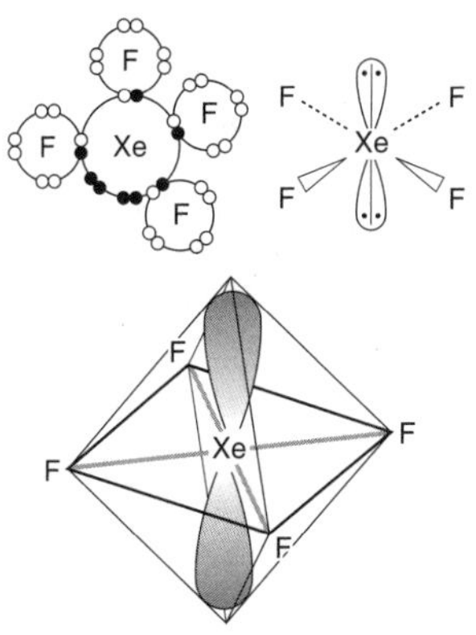

Fig. 2.6

Unit 2

TESTS

RECALL TEST

1 Name the particles in the nucleus and give their charges.

____ (2)

2 State the processes that take place in the mass spectrometer.

____

____ (6)

3 Define

a isotopes

____ (2)

b relative atomic mass

____ (2)

c mass number

____ (2)

4 In a mass spectrum, which peak usually indicates the relative molecular mass of a substance: the peak on the left, the tallest peak, or the peak on the right?

____ (1)

5 Name the shapes of these substances: water ________________, ammonia ________________, methane ________________, beryllium chloride ________________, boron trifluoride ________________, sulphur hexafluoride ________________, phosphorus pentachloride ________________. (7)

6 Work out the shape of the ammonium ion (NH_4^+). What must the bond angle be?

____ (2)

7 What is the shape of the hydroxonium ion, H_3O^+?

____ (1)

8 State the bond angles in: methane __________, ammonia __________, water __________, carbon dioxide __________. (4)

9 What must you state (almost) every time you discuss shape of molecules in exams?

____ (1)

(Total 30 marks)

CONCEPT TEST

1 a In the mass spectrometer:

i How are the ions separated according to mass?

____ (1)

ii Why is a vacuum pump used?

____ (1)

iii Calculate the relative atomic mass of silver from the data left.

____ (2)

Abundance	Isotopic mass
25.7	107
24.3	109

b A pure sample of *P*, a possible pollutant, is put through a mass spectrometer. Chemical analysis indicates the *empirical formula* of the substance is CH_2O. There are peaks on the mass spectrum at 15, 28, 45, 60.

i What is the formula mass of *P*?

______________________________ (1)

ii Which particles produced the peaks with these mass/charge ratios?

15 __________

28 __________

45 __________

60 __________ (4)

iii Using the formula mass and the empirical formula, give two possible structures for *P*. (2)

iv Using the information in **ii**, state the structure of *P*. Explain your answer.

______________________________ (2)

2 The mass spectrum of nickel (atomic number 28) produces the data right:

Abundance	Mass number
33.95	58
13.1	60
0.6	61
1.85	62
0.5	64

a Name the particles in the nucleus and state how many of each are found in the Ni-58 nucleus.

____

______________________________ (3)

b Calculate the relative atomic mass of nickel.

____

______________________________ (2)

3 a Explain why $AlCl_3$ is trigonal planar while NH_3 is pyramidal.

____

______________________________ (3)

b On a sheet of paper, draw the shape of these molecules and ions:

i PH_3 **ii** SO_2 **iii** ClO_3^- **iv** BrF_3^- (4)

c Name the shape of SF_6.

______________________________ (1)

d State and explain the different bond angles in CH_4, NH_3, and H_2O.

Bond angles:

CH_4 __________

NH_3 __________

H_2O __________

Explanation for differences:

____

______________________________ (4)

(Total 30 marks)

Unit 3

GROUPS 1 AND 2

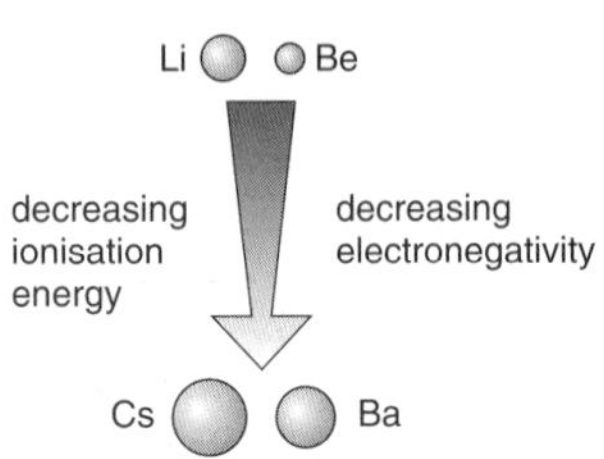

Fig. 3.1

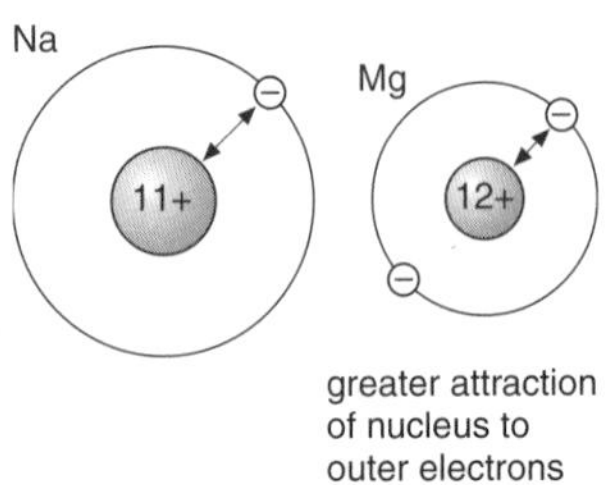

Fig. 3.2

- The elements of groups 1 and 2 make up the **s-block** elements. With increasing atomic number in each group, **metallic bond strength decreases** and the metals become **softer** and have **lower melting points**. Outer bonding electrons become further from the nucleus and are **shielded** from its charge by filled inner shells.

  Similarly, **electronegativity** and **1st ionisation energies** (see unit 5) **decrease** as the electrons in the outer shell become further from the nucleus (see Fig. 3.1).

  For a given period, the group 2 element has **one more proton** in its nucleus and **one more electron** in its outer shell than the group 1 element. Therefore, group 2 metallic bonds are **stronger** than those of group 1 and group 2 metals are harder and have **higher melting points**.

  Similarly, the **electronegativities** and **1st ionisation energies** of the group 2 elements are **higher** than those of the corresponding group 1 elements because the group 2 nuclei have a **greater attraction** for the outer electrons (see Fig. 3.2).

  Group 1 atoms always **lose** control of one electron when they bond: the **oxidation number** in their compounds is always +1. Similarly, the group 2 oxidation number is always +2. (See the section on group 7 for a full explanation of oxidation number).

- **Lithium** and **beryllium** have many properties that are **unlike** those of other members of their groups, e.g. lithium is the only group 1 metal that reacts with nitrogen to form a compound; molten beryllium compounds are poor electrical conductors, indicating strongly covalent character.

- S-block metal compounds usually produce **coloured flames** when strongly heated. Heat **promotes electrons** from lower **energy levels** to higher ones. When the electrons return to lower energy levels, characteristic coloured visible light is emitted, because the energy-level difference is the same as the energy of visible light (see Fig. 3.3).

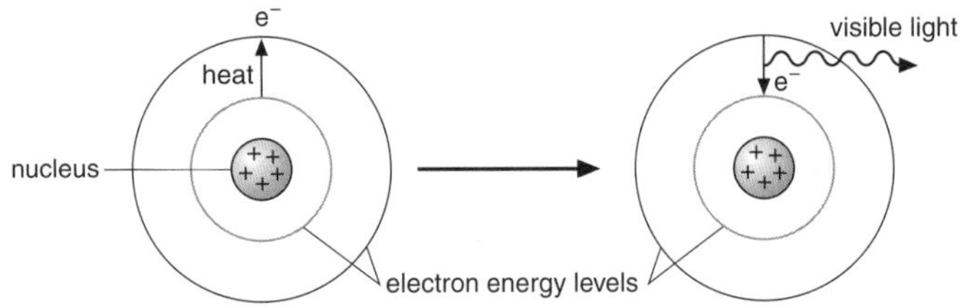

Fig. 3.3

- All s-block compounds are **predominantly ionic** except $BeCl_2$ (see unit 1 for further explanation).

- All s-block elements (except Be) react with chlorine to form **ionic chlorides**. **Examples:**

  Group 1: $2Na(s) + Cl_2(g) \rightarrow 2NaCl(s)$

  Group 2: $Mg(s) + Cl_2(g) \rightarrow MgCl_2(s)$

- All s-block elements react with oxygen to produce **ionic oxides**, e.g.

  Group 1: $4Li(s) + O_2(g) \rightarrow 2Li_2O(s)$

  also $2Na(s) + O_2(g) \rightarrow Na_2O_2(s)$ as well as Na_2O (see opposite)

  Group 2: $2Mg(s) + O_2(g) \rightarrow 2MgO(s)$

  NB **BeO is amphoteric** (like Al_2O_3) showing that Be has non-metallic character.

- In general, s-block elements (not Be) **react with water** to produce hydrogen gas and a metal hydroxide, though some form an insoluble metal oxide (e.g. MgO).

  Group 1: $2Na(s) + 2H_2O(l) \rightarrow 2NaOH(aq) + H_2(g)$

  Group 2: $Ca(s) + 2H_2O(l) \rightarrow Ca(OH)_2(aq) + H_2(g)$

  in **steam**, $Mg(s) + H_2O(g) \rightarrow MgO(s) + H_2(g)$

Flame colours
Li red (crimson)
Na yellow
Ca orange-red
Sr red (crimson)
Ba pale green (apple green)
K lilac or pale purple*
Mg compounds **do not** produce a coloured flame

*You should state that the lilac colour is visible through blue glass.

- You must recall the **acid–base reactions** (even though not all syllabuses state this).

  Metal + Acid → Salt + Hydrogen

  e.g. $Mg(s) + H_2SO_4(aq) \rightarrow MgSO_4(aq) + H_2(g)$

  Metal oxide + Acid → Salt + Water

  e.g. $MgO(s) + H_2SO_4(aq) \rightarrow MgSO_4(aq) + H_2O(l)$

  Metal hydroxide + Acid → Salt + Water

  e.g. $Mg(OH)_2(s) + H_2SO_4(aq) \rightarrow MgSO_4(aq) + 2H_2O(l)$

  Metal carbonate + Acid → Salt + Water + Carbon dioxide

  e.g. $MgCO_3(s) + H_2SO_4(aq) \rightarrow MgSO_4(aq) + H_2O(l) + CO_2(g)$

Amphoteric substances react with both acids and bases.

- Reaction of group 1 metals with **oxygen** can form **simple oxides** (O^{2-} ions), **peroxides** (O_2^{2-} ions), and **superoxides** (O_2^- ions). Peroxides and superoxides are destabilised by small cations; stability increases with increasing cation size. See right.

The Group 1 oxides

Li_2O

Na_2O, Na_2O_2

K_2O, K_2O_2, KO_2

(Rb and Cs as K)

- Generally, an ionic solid is **insoluble** if the energy required to separate the ions from their lattice is much more than the energy released when the same ions are surrounded by water (i.e. are **hydrated**). This means that small (endothermic) **lattice energy** and large (exothermic) **hydration energy** indicate good solubility.
- The **solubilities of group 2 sulphates decrease** with increasing atomic number.

  Barium sulphate is extremely insoluble. Barium ions are used to **test** for the presence of aqueous sulphate anions. Mixing acidified barium chloride (or nitrate) solution with an aqueous sulphate solution produces a thick white precipitate. The acid reacts with sulphite or carbonate and prevents them forming white precipitates of barium sulphite or barium carbonate.

  By contrast, the **solubilities of group 2 hydroxides increase** with increasing atomic number.
- All **group 2 carbonates** decompose when heated to form the oxide and carbon dioxide, e.g. $CaCO_3(s) \rightarrow CaO(s) + CO_2(g)$

  Examine the table below. It shows the temperature at which the group 2 carbonates decompose and the size of the ions.

Compound	Decomposition temperature (K)	Cation radius (10^{-1} nm)
$BeCO_3$	370	0.31
$MgCO_3$	810	0.65
$CaCO_3$	1170	0.99
$SrCO_3$	1560	1.13
$BaCO_3$	1630	1.35

Fig. 3.4 Table of carbonate decomposition temperature and ion size

  The **decomposition temperature increases** as the cation size increases. Carbonate stability decreases with **increasing polarisation** by the cation. The cation distorts the electron shells of the anion and draws them towards itself, **increasing covalent character** (see Fig. 3.5).

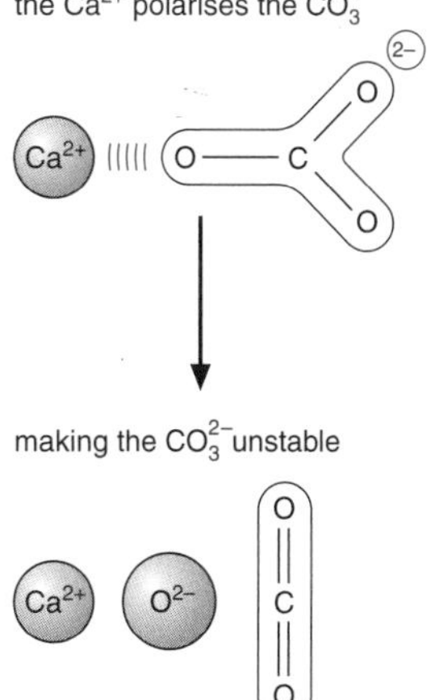

Fig. 3.5

- In general, covalent character is more likely if:
  the charge on the ions is high; AND the cation is small or the anion is large.

  Group 1 carbonates generally do not decompose because the cation charge is only 1+ (and the cations are large), so does not polarise the anion sufficiently. However **Li^+ ions** are very small so they will polarise the carbonate anion:

  $Li_2CO_3(s) \rightarrow Li_2O(s) + CO_2(g)$

Unit 3

TESTS

RECALL TEST

1 Explain why magnesium atoms are smaller than sodium atoms.

____

____ (2)

2 State the group 2 chlorides' flame colours: Mg ____________,
Ca ____________, Sr ____________, Ba ____________,
Li ____________, Na ____________, K ____________. (7)

3 What else must you mention when stating the colour of the potassium flame?

____ (1)

4 Why is barium chloride more ionic than magnesium chloride?

____ (1)

5 Write a balanced equation for each of these reactions:

a magnesium + hydrochloric acid ____________________

b magnesium oxide + hydrochloric acid ____________________

c magnesium carbonate + hydrochloric acid____________________

d magnesium + chlorine ____________________

e magnesium + oxygen ____________________ (5)

6 Why is it difficult to form magnesium peroxide?

____ (2)

7 What is the trend in the solubility of the group 2 sulphates?

____ (1)

8 Explain why the group 2 hydroxides become more soluble with increasing atomic number.

____

____ (3)

9 What is the test for sulphate ions?

____

____ (2)

10 Write an equation for the reaction of limewater with carbon dioxide.

____ (1)

11 Explain why lithium nitrate is unstable to heat.

____ (2)

12 Write an equation for the thermal decomposition of barium carbonate.

____ (1)

13 When sodium peroxide is heated oxygen is given off. Write an equation for this reaction.

____ (1)

14 Barium peroxide and water form hydrogen peroxide. Write an equation for this reaction.

____ (1)

(Total 30 marks)

CONCEPT TEST

1 a When sodium is heated in a Bunsen flame a characteristic yellow flame is seen. Explain why sodium compounds produce coloured flames.

(3)

b Sodium chloride may be made by burning sodium in chlorine. Explain why caesium chloride is not usually made in this way in the laboratory.

(2)

c Write an equation to show the action of water on sodium chloride.

(1)

d Sodium iodide is ionic, while lithium iodide is covalent. Explain why this is so.

(3)

2 Group 2 elements and compounds show some marked trends in physical and chemical properties within the group, with increasing atomic number.

a State and explain the trend in solubility of the group 2 sulphates with increasing atomic number.

(4)

b Aqueous barium ions form a heavy white precipitate with a particular aqueous anion even when acid is added. What is this anion?

(1)

c Explain why the group 2 fluorides increase in solubility with increasing atomic number.

(3)

d Explain why magnesium carbonate decomposes at a moderate temperature but barium carbonate is stable until heated to much higher temperatures.

(2)

e Apply your understanding of group 2 thermal stability to explain why aluminium carbonate decomposes more easily than magnesium carbonate.

(1)

(Total 20 marks)

Unit 4

GROUP 7 AND REDOX

Group 7 elements are also known as **halogens** ('salt generators'); halogen salts are called **halides**.

- You must recall the **colours** and **states** (at room temperature).

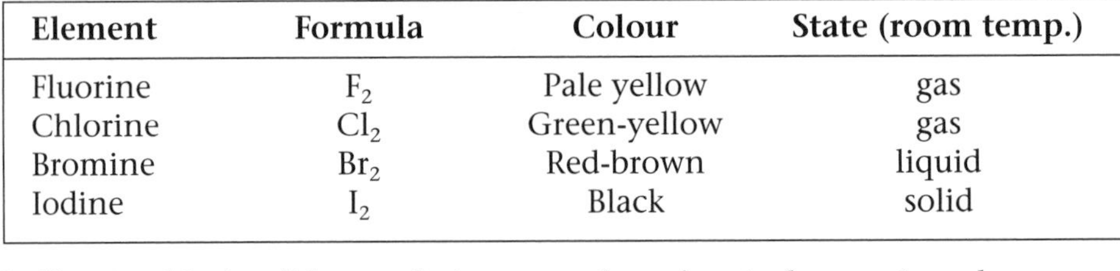

Element	Formula	Colour	State (room temp.)
Fluorine	F_2	Pale yellow	gas
Chlorine	Cl_2	Green-yellow	gas
Bromine	Br_2	Red-brown	liquid
Iodine	I_2	Black	solid

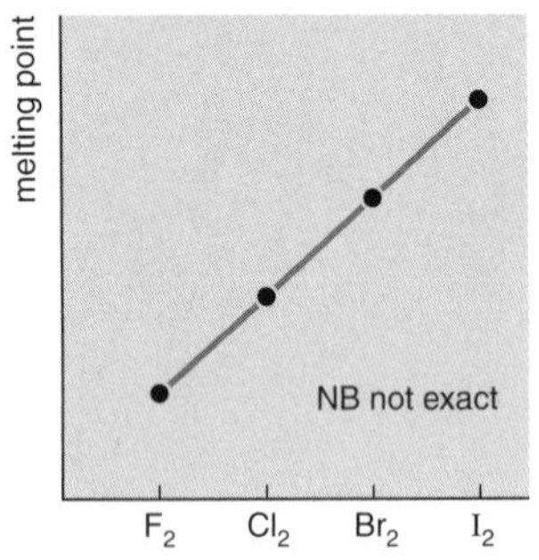

Fig. 4.1 Trend in melting point of halogens

Iodine is a black solid, **purple** in non-polar solvents, **brown** in polar solvents. It is readily soluble in aqueous KI and turns starch blue-black.

As you can see, **melting** and **boiling** points increase with increasing atomic number. This trend is the result of the **Van der Waals forces** increasing with the size of the molecules and the number of electrons present (see Fig. 4.1).

In contrast, **bond strengths** weaken from Cl_2 to I_2 because filled inner shells increase the size of the atoms so that their nuclei are further from the shared (valence) electrons (see Fig. 4.2).

All these elements are strongly attractive to electrons because they each have a high proton number for the period they occupy. The decrease in **electronegativity** and the decrease in **1st ionisation energy** with increasing atomic number are explained in unit 1. (See also unit 5.)

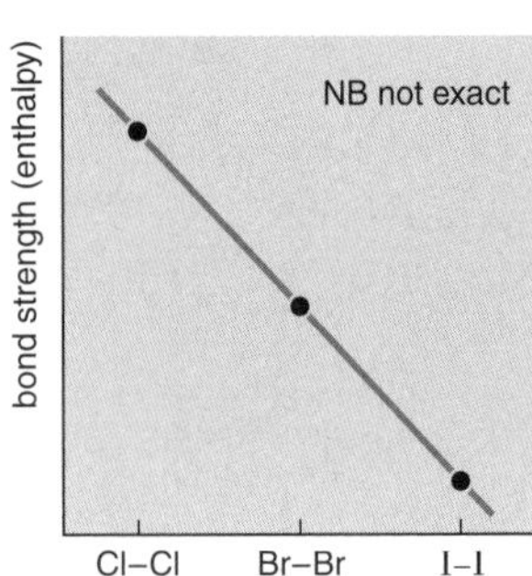

Fig. 4.2

- Chlorine is made industrially by the **electrolysis of brine** (a concentrated solution of impure NaCl) in a membrane (or diaphragm) cell. The **membrane** keeps the electrolysis products separate and stops them from reacting together. Chlorine is produced at the (+) **anode**:

  $2Cl^-(aq) \rightarrow Cl_2(g) + 2e^-$

  The solution passes through the membrane and H_2 is produced at the (–) **cathode**:

  $2H_2O(l) + 2e^- \rightarrow 2OH^-(aq) + H_2(g)$

  Overall: $NaCl(aq) + 2H_2O(l) \rightarrow 2NaOH(aq) + H_2(g) + Cl_2(g)$.

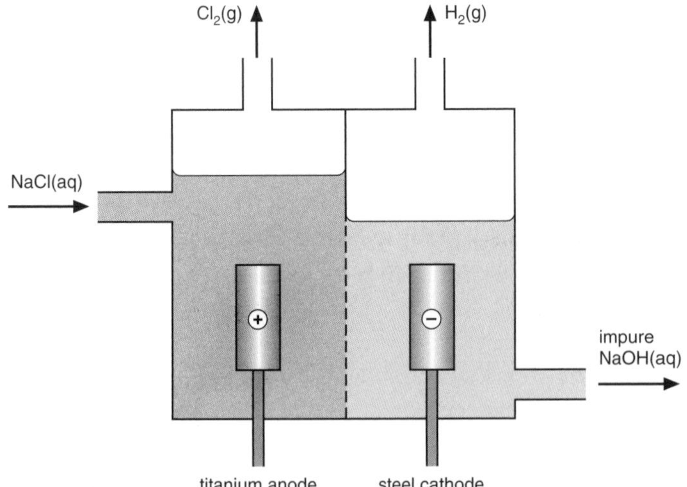

Fig. 4.3 Electrolysis of aqueous NaCl in a diaphragm cell

Sodium hydroxide is extracted from the brine that flows out of the cell (see Fig. 4.3).

All the elements react directly with **metals** to form halides, e.g.

$2Fe(s) + 3Cl_2(g) \rightarrow 2FeCl_3(s)$

Halogen reactivity decreases from F to I as the product lattice or bond energies decrease.

- Halogen/halide **displacement** reactions can happen as the elements compete for electrons. The more reactive halogen will displace a less reactive halogen from a solution containing its ions, e.g.

  $Cl_2(aq) + 2Br^-(aq) \rightarrow Br_2(aq) + 2Cl(aq)$

  $Br_2(aq) + 2Cl^-(aq)$ do not react because Br_2 is not a sufficiently powerful oxidant to oxidise Cl^- ions.

- Chlorine and bromine **bleach** litmus.
- Aqueous **silver nitrate** is used as a **test** for halide ions, forming distinctive silver halide precipitates:

  $Ag^+(aq) + Cl^-(aq) \rightarrow AgCl(s)$ (white solid, soluble in dilute $NH_3(aq)$)

  $Ag^+(aq) + Br^-(aq) \rightarrow AgBr(s)$ (cream off-white solid soluble in conc. $NH_3(aq)$)

  $Ag^+(aq) + I^-(aq) \rightarrow AgI(s)$ (pale yellow solid insoluble in aqueous ammonia)

In contrast **AgF** does not form a precipitate due to the large hydration energy of the small F^- ion.

- HCl, HBr, and HI are gases that are very soluble in water, forming strongly acidic solutions, e.g. $HCl(g) + (aq) \rightarrow H^+(aq) + Cl^-(aq)$.
- **Hydrogen chloride** is produced by mixing (non-volatile) concentrated sulphuric acid with an ionic chloride, e.g.

  $NaCl(s) + H_2SO_4(l) \rightarrow NaHSO_4(aq) + HCl(g)$ (similarly with KF)

  If kept cold KBr will produce **hydrogen bromide**; if hot, then the HBr will reduce the sulphuric acid (an oxidising acid when concentrated) to SO_2:

  $2HBr(g) + H_2SO_4(l) \rightarrow Br_2(g) + SO_2(g) + 2H_2O(l)$

  Hydrogen iodide will further reduce sulphuric acid to form I_2, S, H_2S, and H_2O.
- Chlorine dissolves in water to make **chloric(I) acid**, HClO:

  $Cl_2(g) + H_2O(l) \rightleftharpoons HCl(aq) + HClO(aq)$

  Note that the oxidation number of chlorine is 0 in the Cl_2 and in the products is –1 in HCl and +1 in the **chlorate(I)** ion ClO^-. This simultaneous oxidation and reduction of the same element in a reaction (see below) is called **disproportionation**. It is the HClO in aqueous chlorine that bleaches colour and kills bacteria in drinking water and swimming pools.

  Commercial **bleach** is made by dissolving chlorine in cold aqueous NaOH:

  $Cl_2(g) + 2NaOH(aq) \rightarrow NaCl(aq) + NaClO(aq) + H_2O(l)$

  If warmed the chlorate(I) further disproportionates to give **chlorate(V)**:

  $3ClO^-(aq) \rightarrow ClO_3^-(aq) + 2Cl^-(aq)$

  NaClO and potassium iodide produce brown iodine.
- To determine the amount of iodine in solution, sodium thiosulphate is used with starch as an indicator to help show the presence of iodine:

  $I_2(\text{in } KI(aq)) + 2S_2O_3^{2-}(aq) \rightarrow 2I^-(aq) + S_4O_6^{2-}(aq)$
- **Bromine** is extracted from **sea water** by using chlorine to oxidise bromide ions. The concentration of bromine is increased by absorption in aqueous SO_2. Re-oxidation by chlorine produces sufficient bromine vapour for condensation to a liquid (see Fig. 4.4).

  The main source of **iodine** is sodium iodate(V) which is in the mineral Chile saltpetre, sodium nitrate. Some species of seaweed extract iodine from seawater. Their ash contains up to 0.5% of iodine.

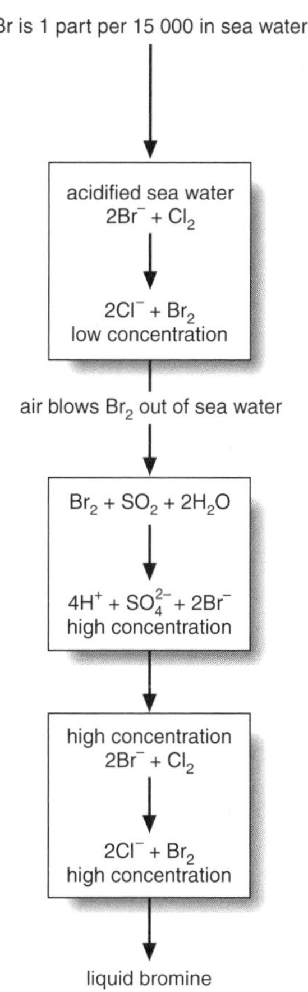

Fig. 4.4 The extraction of bromine from sea water

Redox reactions involve electron transfer. **O**xidation **I**s the **L**oss of electrons; **R**eduction **I**s the **G**ain of electrons. **OILRIG**.

- An **oxidising agent** (oxidant) is a substance that oxidises another substance and is itself reduced in the process. An oxidising agent takes electrons from another substance which acts as a **reducing agent** (reductant).
- **Oxidation number** is the number of electrons an atom has gained or lost control of as a result of its bonding, e.g. when sodium atoms react, they lose one electron per atom so the oxidation number is +1 (note that charge is written the opposite way as 1+).

  Half equations are chemical equations which show the redox change for one substance, with electrons added to balance the equation, e.g. the reaction between chlorate(I) ions and iodide ions in acidified solution to form iodine, water, and chloride ions. The overall redox reaction is written by combining the two half equations, first checking that the numbers of electrons balance. Add the two equations together, cancel the electrons, and add state symbols:

$$ClO^- + 2H^+ + 2e^- \rightarrow Cl^- + H_2O$$

$$2I^- \rightarrow I_2 + 2e^-$$

$$\mathbf{ClO^-(aq) + 2H^+(aq) + 2I^-(aq) \rightarrow Cl^-(aq) + H_2O(l) + I_2(aq)}$$

- Some atoms always have the same oxidation number in a **compound**, e.g.
  Group 1 compounds are always +1, e.g. in NaCl the Na is +1 (and Cl is –1).
  Group 2 compounds are always +2, e.g. in MgO the Mg is +2 (and O is –2).
  Group 3 compounds are usually +3, e.g. in Al_2Cl_6 the Al is +3.
  Fluorine is always –1.
  Oxygen is usually –2 (except when with F or in peroxides), e.g. in MgO the O is –2; in OF_2 the O is +2; and in peroxides, e.g. H_2O_2, the O is –1.
  Chlorine is usually –1 (except when with F or O), e.g. in NaCl the Cl is –1; in ClF_3 the Cl is +3; and in NaClO the Cl is +1.
  Hydrogen is usually +1 (except when it is alone with a less electronegative metal), e.g. in HCl the H is +1; but in NaH the H is –1.
  Some metals have one oxidation number; zinc is +2, silver is +1 (usually). Otherwise you will see the **oxidation state** in Roman numerals after the element, e.g. in iron(II) sulphate Fe has oxidation number +2.
- Remember these **rules** to work out the oxidation number of an **element in a formula**:
  1. The sum of the oxidation numbers in a **compound** always equals zero, e.g. magnesium chloride $MgCl_2$: Mg = +2; Cl = –1; + 2 + (2 × –1) = 0
  2. The sum of the oxidation numbers in an **ion** always equals the charge, e.g. chlorate(I) ClO^-: Cl = +1; O = –2; +1 + (–2) = –1.
  3. The oxidation number of an element in its elemental state always equals zero.

TESTS

RECALL TEST

1 State the colours of the halogens F_2 __________, Cl_2 __________, Br_2 __________, I_2 __________. (2)

2 Why is iodine a solid whereas chlorine is a gas?

____________________ (2)

3 Why is the Cl-Cl bond strong relative to the I-I bond?

____________________ (2)

4 Why is chlorine more reactive than iodine?

____

____________________ (2)

5 Write balanced equations for the reaction of bromine with:

a iron ____________________

b hydrogen ____________________ (2)

6 Write the names of the compounds made when NaCl is added to concentrated sulphuric acid.

____________________ (2)

7 State the products made when KI is added to concentrated sulphuric acid.

____________________ (3)

8 State the oxidation number of the elements in bold beneath the following formulae:

MgO **S**O_2 H_2**S**O_3 **S**O_3 H_2**S**O_4 Mg**S**O_4 H_2**S** **N**H_3 **N**H_4^+ **Na**(s) **Cl**$_2$(g)

____________________ (11)

9 Write equations for the following reactions:

a $Br^-(aq) + Cl_2(aq)$ ____________________

b $Cl^-(aq) + I_2(s)$ ____________________

c $Ag^+(aq) + Cl^-(aq)$ ____________________

d $PCl_5(s) + H_2O(l)$ ____________________ (4)

(Total 30 marks)

CONCEPT TEST

1 Sodium chlorate(VII) has been used for many years as a weedkiller. It is made in three stages, I, II, and III. I: chlorine is made by electrolysis in a diaphragm cell. II: the chlorine is reacted with warm NaOH when disproportionation occurs. III: $NaClO_3$ is dried by warming to make $NaClO_4(s)$.

a In the diaphram cell in stage I,

i identify the electrolyte,

____________________ (1)

ii write the equation for the reaction at the (+) anode.

____________________ (2)

b What is meant by the word 'disproportionation'?

____________________ (2)

c Write the ionic equation for the reaction between warm sodium hydroxide and chlorine. For each chlorine-containing species state the oxidation state of the chlorine.

____________________ (4)

d The $NaClO_4$ forms according to this equation:

$4ClO_3 \rightarrow 3ClO_4^- + 2Cl^-$

Deduce the ionic half equation for the oxidation reaction (assuming alkali is still present).

____________________ (2)

e What is the test for the chloride ions?

____________________ (2)

2 The amount of iodine in a sample may be determined by titrating with aqueous thiosulphate.

a Write the equation for the titration reaction.

____________________ (2)

b Why must starch indicator be used?

____________________ (2)

c How could the presence of iodide be detected?

____________________ (1)

d When chlorine gas is bubbled through potassium iodide solution iodine forms.

State two changes that could be observed.

____________________ (2)

(Total 20 marks)

PERIODIC TABLE I: PHYSICAL

> There are different types of atomic orbital (s, p, d, f), but you only have to know the shapes of **s** and **p** orbitals.

- The space occupied by an electron around a nucleus for 95% of the time is called an **atomic orbital**. Up to two electrons can occupy each atomic orbital.

  s orbitals are spherical with the nucleus at the centre (see Fig. 5.1).
  p orbitals are shaped like two balls stuck together, with the nucleus where the spheres touch (see Fig. 5.2).

  Orbitals of the same type group together in **subshells**. There is one s orbital in a s subshell, three p orbitals in a p subshell, and five d orbitals in a d subshell.

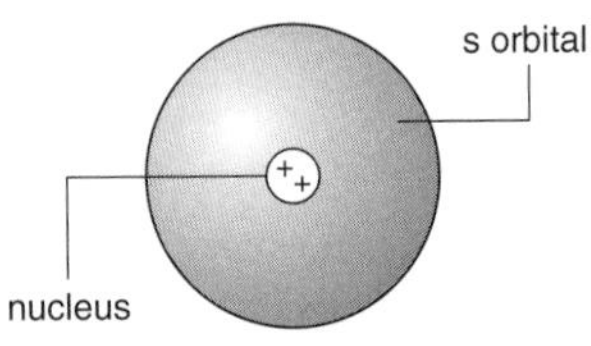

Fig. 5.1 An s orbital

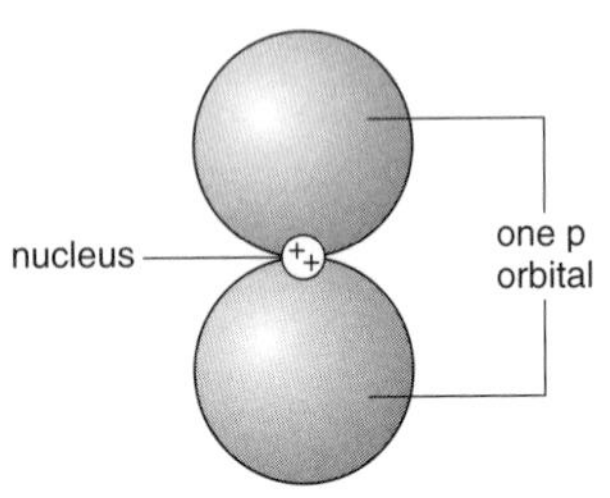

Fig. 5.2 A p orbital

- **Shells** contain groups of subshells that have similar energies. In a given shell, the energies of the subshells increase in the order s < p < d (see Fig. 5.3).

  A list of an atom's occupied subshells is called its **electronic configuration**. You may be asked to write the configuration for any of the elements 1 (H) to 36 (Kr), e.g. Kr = $1s^2\ 2s^2\ 2p^6\ 3s^2\ 3p^6\ 3d^{10}\ 4s^2\ 4p^6$. The term $3p^6$, for example, means that there are 6 electrons in the p subshell that is part of the shell n = 3. Note that 4s is filled before 3d (see unit 24). Use the periodic table to help you write electronic configurations. The group 1 and 2 metals are in the **s block** because the s orbital is being filled. Similarly, the block on the right (groups 3–0) is called the **p block**. The period (row) number tells you the principle quantum number of the s or p subshell being filled.

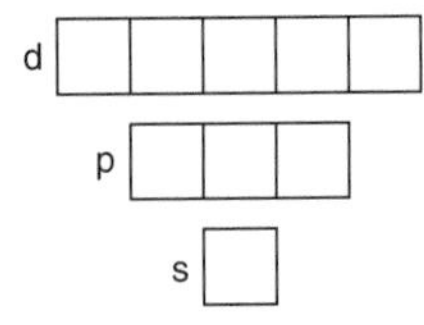

Fig. 5.3

- You must recall that the electronic configuration determines the **chemical properties** of an element.

- **Covalent bonds** form when atomic orbitals overlap to make a **bonding molecular orbital**. Electrons in the molecular orbitals are shared between the atoms.

  A single bond is made when two orbitals overlap end on. This bond is called a **sigma (σ) bond** (see Fig. 5.4).

  Two p orbitals overlap sideways to make a single **pi (π) bond** (see Fig. 5.5).

  A **double bond** consists of two bonds – a sigma and a pi bond. The sigma bond is symmetrical around the line joining the two nuclei. The pi bond exists either side of the sigma bond.

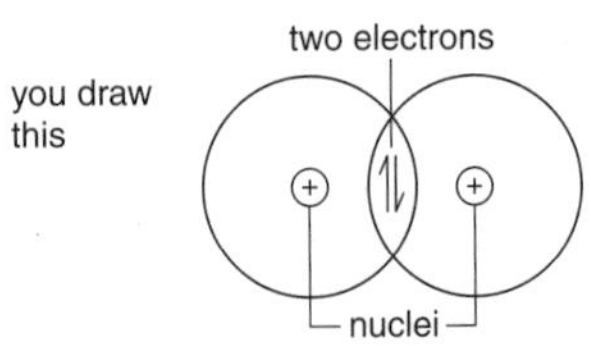

Fig. 5.4 A σ bond

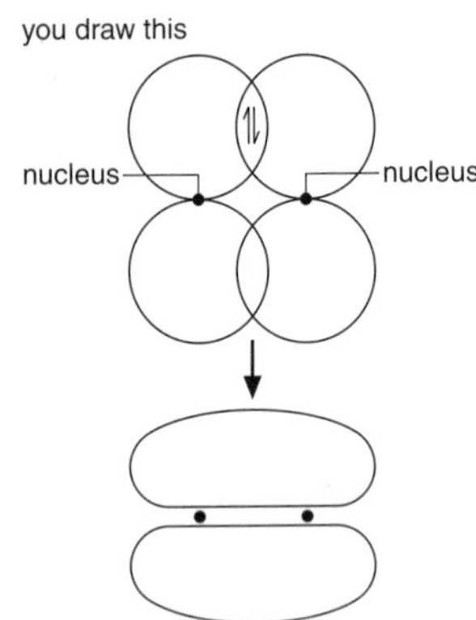

Fig. 5.5 A π bond

- You must recall the definitions:
  1st electron affinity (EA): $X(g) + e^- \rightarrow X^-(g)$ The energy released when one mole of electrons is gained by one mole of gaseous atoms to form one mole of gaseous ions with a single negative charge.
  (**Slightly exothermic** as the nucleus in the neutral atom attracts the electron).
  2nd electron affinity: $X^-(g) + e^- \rightarrow X^{2-}(g)$ (**Greatly endothermic** as the negative ion repels the electron).
  Note the combined 1st and 2nd electron affinity is endothermic.
  1st and 2nd EA: $X(g) + 2e^- \rightarrow X^{2-}(g)$ (see Fig. 5.6.)

The electron affinities of oxygen

1st EA −142 kJ mol^{-1}
$O(g) + e^- \longrightarrow O^-(g)$

2nd EA +844 kJ mol^{-1}
$O^-(g) + e^- \longrightarrow O^{2-}(g)$

Fig. 5.6

- You may have to explain why metal oxides form readily from their elements, even though making the oxide ion from gaseous atoms is highly endothermic (costs energy). The answer is that a huge amount of energy (the lattice formation enthalpy) is released when the oxide ion and the metal ion join together.

- You must recall the definition:
  1st ionisation energy: $X(g) \rightarrow X^+(g) + e^-$ The energy required to remove one mole of electrons from one mole of gaseous atoms to form one mole of gaseous ions with a single positive charge.
  You may have to define successive ionisation energies of an element, for example:
  3rd ionisation energy: $X^{2+}(g) \rightarrow X^{3+}(g) + e^-$ The energy required to remove one mole of electrons from one mole of gaseous X^{2+} ions to form one mole of gaseous X^{3+} ions.

It takes energy to pull the (–) e from the (+) ion, so all these changes are **endothermic**. As successive electrons are removed the cation charge increases so the ionisation energies increase.

- Examine the graph of the 1st ionisation energies of **successive elements** (Fig. 5.7).

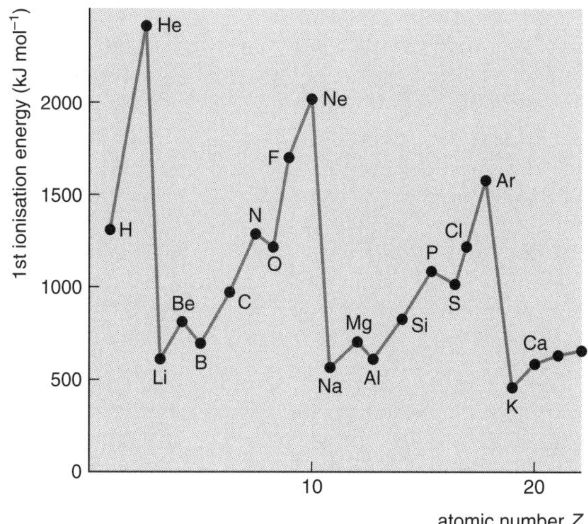

Fig. 5.7

Note:
(1) the general **increase** in 1st ionisation energy **from Na** to **Ar**, due to increased proton number.
(2) the **peaks** in 1st ionisation energy are always the **noble gases** because these have the highest proton number for the period, just before a new shell starts in the next period.
(3) there is a **drop** in 1st ionisation energy between **group 2 and 3** elements. The electron in the group 3 atom is lost more easily from the p orbital, which is further from the nucleus, than the electron lost from the s orbital in the group 2 atom.
(4) there is a **drop** between **groups 5 and 6**. Group 6 atoms have two electrons paired in the p subshell. Repulsion between these electrons makes one of them easier to remove.

- The elements are arranged in the periodic table in order of **increasing proton number** (atomic number). The periods (rows) show **repeating** physical and chemical properties (this is known as **periodicity**). The elements in a given group (column) have similar properties because the outer shells of electrons have similar structures.

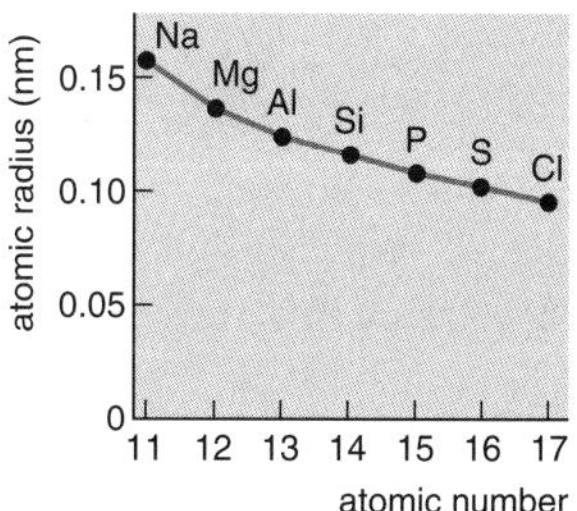

Fig. 5.8

The **atomic radius** decreases across a period as the proton number increases and pulls in the outer electrons (see Fig. 5.8).

For the same reasons, the **cation radius** decreases as the nuclear charge increases, while the electronic configurations of the common ions in a period are similar (see Fig. 5.9).

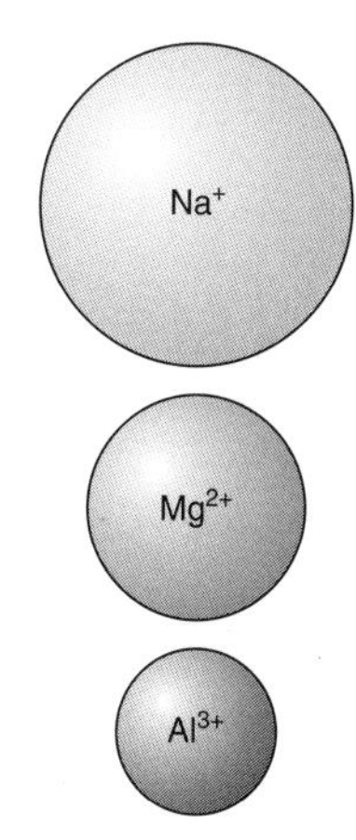

Fig. 5.9 The relative sizes of the Na^+, Mg^{2+}, and Al^{3+} ions

The **anions** in a period have one more complete shell than the cations, making the anions larger, while the **anionic radius** decreases as nuclear charge increases.

The proton number increases **across a period** (row) and so the nucleus becomes more attractive to electrons. This trend explains the increases in electronegativity, 1st ionisation energy, and 1st electron affinity.

The proton number increases **down a group** and so the number of filled electron shells increases, which (i) increases the distance between the nucleus and the outer electrons and (ii) increases shielding of the nuclear charge. The atomic radius increases while the nuclear attraction for the outer electrons decreases, causing electronegativity and 1st ionisation energy to decrease.

TESTS

RECALL TEST

1 What is meant by an 'atomic orbital'?

______________________________ (2)

2 On a piece of paper, draw

a a s orbital (1)

b a p orbital (1)

c a sigma bond (1)

d a pi bond (1)

3 Write the electronic configuration of krypton.

______________________________ (2)

4 What is meant by 's-block element'?

______________________________ (1)

5 State what determines the chemical properties of an element.

______________________________ (1)

6 Why does electronegativity increase across the third row of the periodic table?

______________________________ (1)

7 State and explain the change in 1st ionisation energy across the third row of the periodic table.

______________________________ (1)

8 In group 2, with increasing atomic number, state whether the follow increase, decrease, or stay the same:

a the atomic radii ______________ (1)

b electronegativity ______________ (1)

c 1st ionisation energy ______________ (1)

9 Define '1st electron affinity'.

______________________________ (1)

10 Why is the 1st electron affinity only slightly exothermic whereas the 2nd electron affinity is a large endothermic energy change?

______________________________ (2)

11 Define 4th ionisation energy.

______________________________ (2)

12 Which of the first 20 elements (H to Ca) has:

a the highest 1st ionisation energy ______________ (1)

b the weakest Van der Waals forces ______________ (1)

c the smallest cationic radius ______________ (1)

d the smallest anionic radius ______________ (1)

13 For the third period explain why the melting point of the elements starts high (for Na, Mg, Al) then peaks at Si, but is lower for P, S, Cl, Ar.

______________________________ (6)

(Total 30 marks)

CONCEPT TEST

1 This question concentrates on ionisation energy.

a Give an equation for the first ionisation energy of chlorine atoms.

____ (2)

b The first ionisation energy of helium is the highest of all atoms. Explain why this is so.

____ (2)

c Give equations that represent the first electron affinity and second electron affinity of sulphur.

1st electron affinity of S:

____

2nd electron affinity of S:

____ (2)

d Metal sulphides form from their elements even though the combined first and second electron affinity of sulphur is endothermic. Explain why metal sulphides are so common.

____

____

____

____ (3)

2 State which element has the higher ionisation energy (IE) for each pair of elements. Give a short explanation in each case.

a 1st IE of C and 1st IE of Si. ______________________

b 1st IE of Ar and 1st IE of K. ______________________

c 1st IE of Be and 1st IE of B. ______________________

d 1st IE of Na and 1st IE of Mg. ______________________

e 2nd IE of Na and 1st IE of Mg. ______________________ (10)

3 a On a piece of paper, draw a graph of the log of successive ionisation energies of potassium. (4)

b Explain why the successive ionisation energies of potassium generally increase.

____ (1)

c Explain the shape of the graph.

____

____ (4)

d Why are there only 19 ionisation energies for potassium?

____ (2)

(Total 30 marks)

Unit 6

PERIODIC TABLE II: CHEMICAL

- Knowing the **oxidation states** will help you to work out the **formulae** of compounds. Knowing the **electronegativity** values will help you deduce the type of **bonding** and **structure** present in different compounds. In turn, this knowledge will help you to recall their **chemical properties** and reactions.
- You need to know the formulae, bonding, and structure of the **third period** elements (Na–Ar) and their compounds (see below). You could use the oxidation states to help you work out the formulae. Remember to state that argon is so inert that it does not form compounds.

Elements	Oxides	Chlorides	
Na	Na_2O, Na_2O_2	NaCl	NaOH
Mg	MgO	$MgCl_2$	$Mg(OH)_2$
Al	Al_2O_3	Al_2Cl_6	$Al(OH)_3$
Si	SiO_2	$SiCl_4$	$Si(OH)_4$
P	P_4O_6, P_4O_{10}	PCl_3, PCl_5	H_3PO_3, H_3PO_4
S	SO_2, SO_3	S_2Cl_2	H_2SO_3, H_2SO_4
Cl	Cl_2O	Cl_2	HClO, $HClO_4$
Ar	none	none	

- Most elements burn in oxygen to form oxides. (See below for Si.)

Reaction with oxygen	Oxidation number of element	Product bonding and structure
$2Na(s) + O_2(g) \rightarrow 2Na_2O_2(s)$ or $Na_2O(s)$	+1	ionic lattice
$2Mg(s) + O_2(g) \rightarrow 2MgO(s)$	+2	ionic lattice
$2Al(s) + 3O_2(g) \rightarrow 2Al_2O_3(s)$	+3	giant ionic/covalent lattice
$4P(s) + 5O_2(g) \rightarrow P_4O_{10}(s)$	+5	simple covalent
$S(s) + O_2(g) \rightarrow SO_2(g)$	+4	simple covalent

Silicon dioxide has a giant covalent lattice and so has a very high melting point. It is chemically rather inert. **Solid silicon** does not react easily as it is coated with a layer of silicon dioxide. **Aluminium** also has an impervious oxide layer which stops it reacting easily with water and oxygen.

- The elements react with **chlorine** to form chlorides.

Reaction with chlorine	Oxidation number of element	Product bonding and structure
$2Na(s) + Cl_2(g) \rightarrow 2NaCl(s)$	+1	ionic lattice
$Mg(s) + Cl_2(g) \rightarrow MgCl_2(s)$	+2	ionic lattice
$2Al(s) + 3Cl_2 \rightarrow Al_2Cl_6(s)$	+3	simple covalent
$Si(s) + 2Cl_2(g) \rightarrow SiCl_4(l)$	+4	simple covalent
$2P(s) + 5Cl_2(g) \rightarrow 2PCl_5(s)$	+3	simple covalent
$2S(s) + Cl_2(g) \rightarrow S_2Cl_2(l)$	+1	simple covalent

- Only sodium, magnesium, and chlorine react with **water**.

$2Na(s) + 2H_2O \rightarrow 2NaOH(aq) + H_2(g)$
(vigorously) oxidation number +1 pH 12–14

$Mg(s) + H_2O(l) \rightarrow MgO(s) + H_2(g)$
(Mg only burns in steam) oxidation number +2 pH 8–9 (in water)

In the periodic table, there are many **patterns** that can help you to learn the chemistry of the elements.

The **oxidation states** show clear trends and patterns: learn them.

Iron is extracted in a blast furnace. The raw materials are iron ore (haematite), coke (carbon), and limestone. Hot air is blown into the furnace to burn C (producing CO_2) and to heat the furnace. Some CO_2 reacts with hot C to make CO, which reduces the Fe_2O_3 to Fe. Limestone reacts with acid impurities to form slag that floats on the molten iron.

Aluminium is made from purified **bauxite**. To **electrolyse** pure Al_2O_3 it is dissolved in molten cryolite. At graphite cathode (–):

$Al^{3+} + 3e^- \rightarrow Al(l)$

At graphite anode (+):

$2O^{2} \rightarrow O_2 + 4e^-$

then

$C(s) + O_2(g) \rightarrow CO_2(g)$

This process is expensive because of its high energy cost. Recycling only uses a small amount of energy.

- Some **oxides** dissolve to give acidic or alkaline **solutions**. Note that the oxidation number of the element does not change.

Oxide with water	Oxidation number of element	pH	Oxide type
$Na_2O(aq) + H_2O(l) \rightarrow 2NaOH(aq)$	+1	12–14	basic
$MgO(s) + H_2O(l) \rightarrow Mg(OH)_2(aq)$	+2	8–9	basic
$Al_2O_3(s) + H_2O(l) \rightarrow$ no reaction	+3	7	amphoteric
$SiO_2(s) + H_2O(l) \rightarrow$ no reaction	+4	7	acidic
$P_4O_{10}(s) + 6H_2O(l) \rightarrow 4H_3PO_4$*	+5	1	acidic
$SO_2(g) + H_2O(l) \rightleftharpoons H_2SO_3(aq)$	+4	3	acidic
$SO_3(g) + H_2O(l) \rightarrow H_2SO_4$** (reacts violently to make strong acid)	+6	1	acidic

* Either (s) if pure or (aq) if dilute.

** Either (l) if pure or (aq) if dilute.

- The **ionic chlorides** dissolve in water:

Chloride with water	pH
$NaCl(s) + (aq) \rightarrow Na^+(aq) + Cl^-(aq)$	pH 7
$MgCl_2(s) + (aq) \rightarrow Mg^{2+}(aq) + 2Cl^-(aq)$	pH 6

- The **covalent chlorides** vigorously hydrolyse in water, producing white fumes of HCl:

Chloride with water	pH
$Al_2Cl_6(s) + 6H_2O(l) \rightarrow 2Al(OH)(s) + 6HCl(g)$	pH 1 (due to the HCl)
$SiCl_4(l) + 4H_2O(l) \rightarrow Si(OH)_4(s) + 4HCl(g)$	pH 1 (due to the HCl)
$PCl_5(s) + 4H_2O(l) \rightarrow H_3PO_4(aq) + 5HCl(g)$	pH 1 (due to the HCl and H_3PO_4)
$2S_2Cl_2(l) + 2H_2O(l) \rightarrow 3S(s) + SO_2(g) + 4HCl(g)$	pH 1 (a slower redox reaction)

- Some chlorides, such as **carbon tetrachloride**, do not dissolve in water or react with it. When water reacts with a simple covalent chloride, such as $SiCl_4$, the water molecule donates a lone pair of electrons into the **low energy vacant orbitals** in the outer shell of the Si atom. The water molecules join with the silicon and an H^+ ion leaves to join with the Cl^- to make HCl (See Fig. 6.1). CCl_4 has no low-energy vacant orbitals in its outer shell so water does not react with it. In addition C-Cl bonds are stronger than Si-Si bonds.

Generally, the covalent chlorides produce the **same acid** that the oxides produce in water.

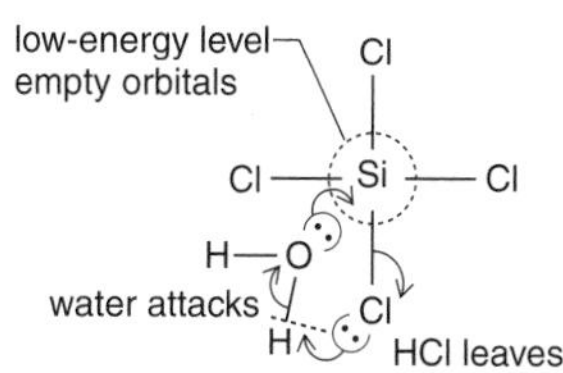

Fig. 6.1

- Most **simple covalent chlorides** exist as covalent molecules; some are covalent molecules when gases, but are **partially ionic** when solids or liquids.
- The **group 4 elements** (C, Si, Ge, Sn, Pb) show the **clearest trend** from non-metallic to metallic properties as atomic number increases. As the number of filled electron shells increases, nuclear attraction for the outer electrons weakens and the elements become more metallic. (See unit 1.)

  Carbon, **silicon**, and **germanium** have **giant covalent structures** with strong bonds that result in high melting points. Shielding increases from C to Si, which causes the covalent bonds to weaken and melting points to fall. **Tin** and **lead** have **metallic structures**; the relatively weaker metallic bonding gives lower melting points. The large size of Sn and Pb atoms result in **lower melting points** than many of the common metals with smaller metallic radii.

  The **+4 oxidation state** becomes **less stable** down the group C to Pb and the +2 state becomes more stable. As a result, Sn(II) is a strong reducing agent, that readily oxidizes from the +2 state to its preferred state of +4. Conversely, Pb tends to reduce from +4 to +2, making Pb(IV) a strong oxidising agent.

Unit 6

TESTS

RECALL TEST

1 State the formulae of the elements, oxides, and chlorides of the third period (Na to Ar) by filling in the table below:

Element	Na	Mg	Al	Si	_	_	_	Ar
Oxide								
Chloride								

(16)

2 Group the elements and compounds in the table by structure and bonding. (There are four types of structure in question: metallic, ionic, giant covalent lattice, and simple covalent molecules.)

(2)

3 Write balanced equations for the reactions of the elements Na to S with oxygen.

(6)

4 Write balanced equations for the reactions of the elements Na to S with chlorine.

(6)

5 Write balanced equations for the reactions of the few elements that react with water, and state the pH.

(3)

6 Write balanced equations for the reactions of the oxides with water, and suggest a pH for each solution formed.

(10)

7 Write balanced equations for the action of water on the chlorides NaCl to PCl_5.

(5)

8 Why does silicon tetrachloride react with water, while CCl_4 does not?

(2)

(Total 50 marks)

CONCEPT TEST

1 This question concentrates on oxides. Fill in the gaps in this table:

Element	Mg	Al	Si	S
Oxide				
Formula(e)				
Bonding				
Structure				

(9)

a Why is it difficult to prepare SiO_2 from silicon in the laboratory?

____

______________________________ (2)

b Write a balanced equation for:

i making magnesium oxide from its constituent elements,

____

ii reacting phosphorus(III) oxide with water,

____

iii reacting sulphur(IV) oxide with water.

______________________________ (3)

2 This question concerns the chlorides. Fill in the table:

Element	Na	Al	Si	P
Chloride				
Formula(e)				
Bonding				
Structure				

(9)

a Describe how phosphorus(III) chloride may be prepared in the laboratory.

____

______________________________ (3)

b Give balanced equations for the reaction of chlorine with:

i phosphorus,

____

ii sodium.

______________________________ (2)

c Give balanced equations for the following (or state there is no reaction):

i the action of water on sodium chloride,

____

ii the action of water on silicon chloride.

______________________________ (2)

(Total 30 marks)

ORGANIC BONDING AND ISOMERS

- Organic compounds are based on **skeletons** of **carbon atoms** covalently bonded together in **chains** and **rings**. (See unit 1.) When different organic molecules have similar structures and react similarly, then we list them in a group called a **homologous series**. The members of each series have similar structures but different numbers of -CH_2- groups.
- The members of the **alkane** homologous series consist of only carbon and hydrogen atoms joined by single covalent bonds. You need to recall the first ten alkane names and formulae. (The general formula C_nH_{2n+2} will help).

Number of C atoms	Name	Formula
1	methane	CH_4
2	ethane	C_2H_6
3	propane	C_3H_8
4	butane	C_4H_{10}
5	pentane	C_5H_{12}
6	hexane	C_6H_{14}
7	heptane	C_7H_{16}
8	octane	C_8H_{18}
9	nonane	C_9H_{20}
10	decane	$C_{10}H_{22}$

The **boiling points** and the **melting points** of the alkanes increase with increasing number of carbon atoms.

Straight chain molecules have higher boiling points than **branched isomers** of the same size. **Example:** pentane C_5H_{12} b.p. = 36 °C; methylbutane C_5H_{12} b.p. = 28 °C.

The **alkanes** are unreactive because they contain strong C-C and C-H bonds. The **bonds** are **strong** (have a high average bond enthalpy – see unit 12) because the atoms are very small and the outermost electrons are not shielded from the attraction of the nuclear charge. This effect causes the C-C bond, for example, to be **short** and strong (see Fig. 7.1).

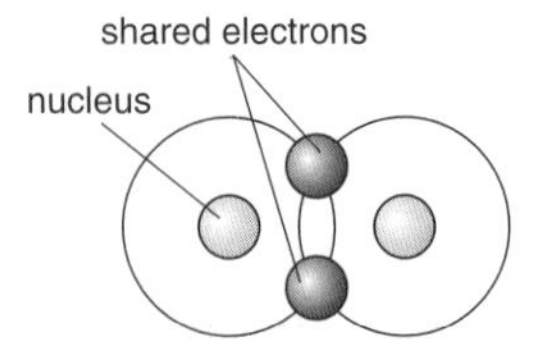
shared electrons
nucleus
The nucleus is very close to the shared electrons so the opposite charges attract strongly.

Fig. 7.1

The **alkenes** contain C=C double bonds. They are a more reactive series of compounds than the alkanes. A double bond consists of a strong sigma bond and a weaker pi bond (see unit 1). Electrons in the sigma bond are concentrated between the two nuclei; electrons in the pi bond are concentrated further away and to the sides of the nuclei, resulting in a weaker bond (see Fig. 7.2).

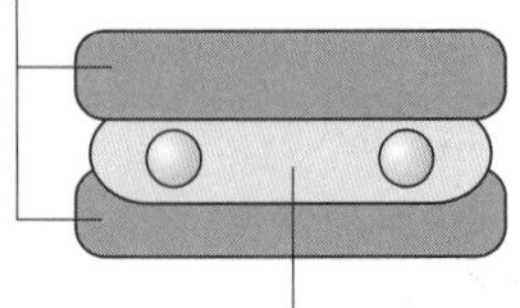
The electrons in the pi bond are far from the nuclei.
The electrons in the sigma bond are close to the nuclei.

Fig. 7.2

- The **halogenoalkanes** are a homologous series of compounds that consist of alkanes which have one or more hydrogen atoms replaced by a halogen atom. Halogen atoms are approximately 2 to 4 times larger than hydrogen atoms. The nucleus of the large halogen atom is far from the shared electrons in the carbon–halogen bond. For example, the bromine atom is so large that the C-Br bond is just two-thirds the strength of the C-H bond (see Fig. 7.3).

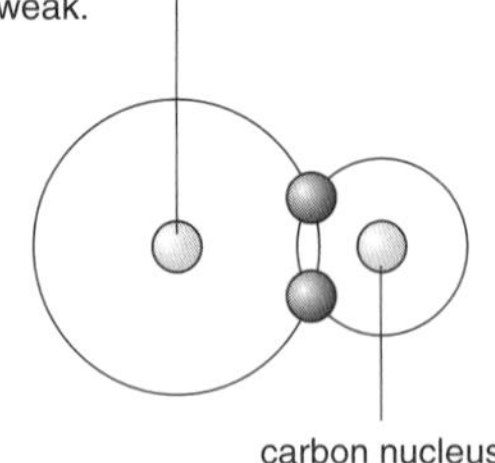
This bromine nucleus is far from the shared electrons so the C–Br bond is weak.
carbon nucleus

Fig. 7.3

Alkanes are so **unreactive** that you must use harsh conditions to make them react. Ultraviolet light has high energy that will break bonds in alkanes and start a reaction (see unit 8 for details). High temperature can also be used to initiate reactions, as in the combustion of petrol (consisting mostly of alkanes).

Alkenes are more **reactive** than alkanes. Their reactions often take place at 'room temperature'.

Some **halogenoalkanes** (chloro-, bromo-, and iodoalkanes) have weaker covalent bonds than alkanes so **react faster**. However, they still need a little help, so the reaction mixture is usually heated.

Organic reactions are usually **slow**. Heating a flask of chemical reagents allows volatile liquids to boil and escape. Stoppering the heated flask would cause an explosion. The solution is to fit a **vertical condenser** to a heated flask. This arrangement allows extended boiling without loss of volatile substances. In exams you need to write this condition as **heat under reflux**.

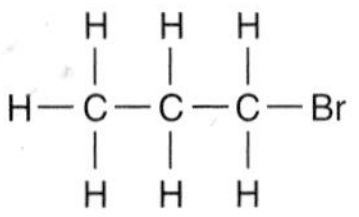

- Look at the structures of 1-bromopropane and 2-bromopropane (see Fig. 7.4). Both have the same numbers of atoms of C, H, and Br, and the same formula C_3H_7Br. These compounds have different structures, so we call them **structural isomers**.

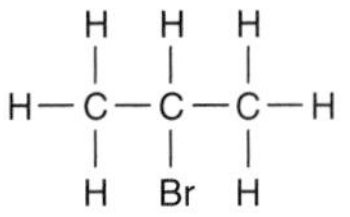

  Some compounds containing **C=C double bonds** exhibit **geometrical isomerism**. Look at the structures of *cis*-but-2-ene and *trans*-but-2-ene (see Fig. 7.5). The atoms are joined together in the same order $CH_3CHCHCH_3$ but the molecules have different structures because the two methyl $-CH_3$ groups are either on the same side (***cis***) or on different sides (***trans***) of the C=C double bond.

  Alkanes do not exhibit geometrical isomerism because the atoms can **rotate** around a single C-C bond. There are no geometrical isomers of butane $CH_3CH_2CH_2CH_3$. **Alkenes** show this type of isomerism because of the **restricted rotation** around the C=C bond.

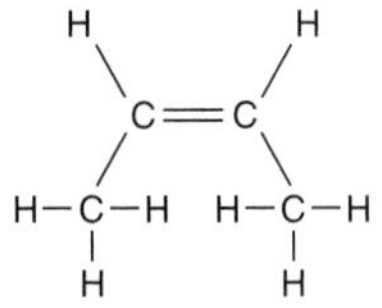

- The presence of C=C and C-Br bonds result in a reactive site on a molecule, so the C=C and C-Br are called **functional groups**. You do need to learn these functional groups as soon as possible. You could start with the top six in the table below. The functional groups are shown in bold type.

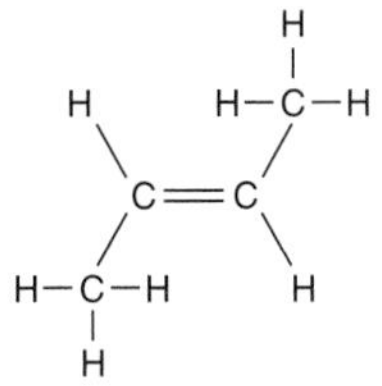

1-bromopropane

2-bromopropane

Fig. 7.4

cis-but-2-ene

trans-but-2-ene

Fig. 7.5

Homologous series	Name of example	Graphical formula	Linear abbreviated formula
alkanes	propane	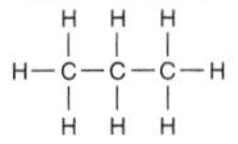	$CH_3CH_2CH_3$
alkenes	propene	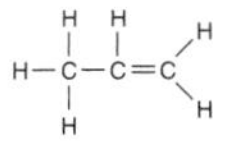	$H_2\mathbf{C{=}C}HCH_3$
halogenoalkanes	1-bromopropane	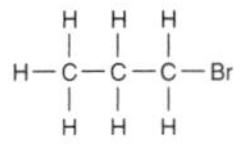	$CH_3CH_2CH_2\mathbf{Br}$
alcohols	propan-1-ol	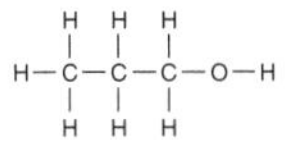	$CH_3CH_2CH_2\mathbf{OH}$
aldehydes	propanal	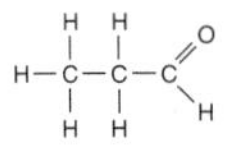	$CH_3CH_2\mathbf{CHO}$
ketones	propanone	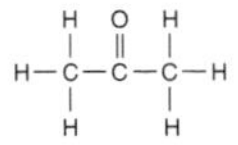	$CH_3\mathbf{CO}CH_3$
amines	1-aminopropane	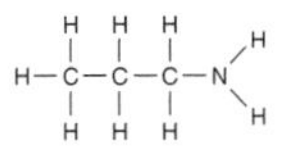	$CH_3CH_2CH_2\mathbf{NH_2}$
nitriles	propanenitrile	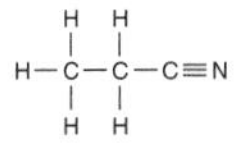	$CH_3CH_2\mathbf{CN}$
carboxylic acids	propanoic acid	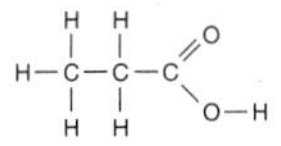	$CH_3CH_2\mathbf{COOH}$
carboxylic acid salts	sodium propanoate	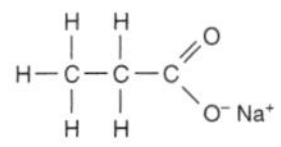	$CH_3CH_2\mathbf{CO_2^-Na^+}$
esters	ethyl propanoate	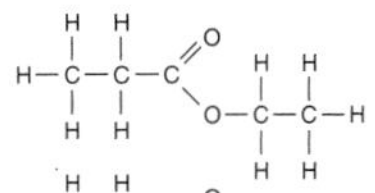	$CH_3CH_2\mathbf{COO}CH_2CH_3$
amides	propanamide	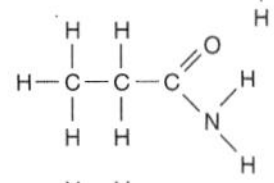	$CH_3CH_2\mathbf{CONH_2}$
carbonyl chlorides	propanoyl chloride	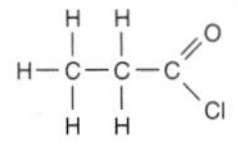	$CH_3CH_2\mathbf{COCl}$

TESTS

RECALL TEST

1 What is a homologous series?

______________________________ (1)

2 Name and write the formulae of the first ten alkanes.

____

______________________________ (3)

3 Why are the alkanes so unreactive?

____

______________________________ (1)

4 Why are alkenes very reactive?

____

______________________________ (1)

5 Why are halogenoalkanes more reactive than alkanes?

____

______________________________ (1)

6 What sort of conditions are required for the following to react?

a alkanes ______________________________

b alkenes ______________________________

c halogenoalkanes ______________________________ (3)

7 Explain

a structural isomerism,

______________________________ (1)

b geometric isomerism.

______________________________ (1)

8 How many structural isomers may be made from C_4H_8? ______________ (1)

9 How many isomers may be made from C_4H_7Br? ______________ (1)

10 In a covalent molecule, how many bonds do these make?

a carbon ____ b nitrogen ____ c oxygen ____ d fluorine ____ (4)

11 State and explain the change in the strength of these covalent bonds: C-F, C-Cl, C-Br, C-I.

____

______________________________ (2)

12 For each of the functional groups in the table on page 27, state the intermolecular force (or bonding) present that determines the boiling point of the substance.

____

____

____

______________________________ (7)

13 On paper, show the graphical (see Fig. 9.2) and linear formula of propan-2-ol and then show how it can be simplified to lines and an OH. (3)

(Total 30 marks)

CONCEPT TEST

1 When a solution of bromine is shaken with hexene, C_6H_{12}, the bromine is decolorised. However, when bromine is added to hexane in the dark there is no decolorisation.

a Write an equation for the reaction of hexene with bromine.

(1)

b Explain, in terms of the bonding, why no reaction occurs when a solution of bromine is shaken with hexane in the dark.

(2)

2 Suggest the conditions required for these three reactions:

a Ethane may be mixed with HBr to form bromoethane.

b Bromoethane will react with NaOH(aq) to form ethanol.

c Bromoethane may also be formed from ethane.

(3)

3 (See also unit 11) Consider these bond enthalpies.

Bond	Enthalpy (kJ mol^{-1})
F-H	562
Cl-H	431
Br-H	366
I-H	299
C-C	348
C=C	612
Si-Si	176

a Explain the trend in the halogen–hydrogen bond enthalpies.

(3)

b Explain the difference between the C-C and C=C bond enthalpies.

(3)

c Explain why the Si-Si bond is weaker than the C-C bond.

(3)

4 The compound right is being considered as an insecticide:

it may be drawn like this:

a On a piece of paper draw the *cis/trans* isomers that this compound may have.

b Explain what makes this molecule so reactive.

(5)

(Total 20 marks)

Unit 8

ORGANIC MECHANISMS

- There are millions of possible organic reactions but most of them fall into one of the ten classes described here. Each type of reaction proceeds from reactants to products in a distinct series of steps, known as the **reaction mechanism**.
- **Acid–base reactions** happen when an organic molecule **donates** a proton H^+ or accepts a proton. (See Fig. 8.1.)

  When a molecule **donates** a proton, then it is acting as an **acid**. An **acidic hydrogen atom** leaves the molecule as H^+ and is replaced by a metal ion (or NH_4^+). **Example:** $CH_3COO\mathbf{H} + Na\mathbf{OH} \rightarrow CH_3COO^-Na^+ + \mathbf{H_2O}$

  Example: $CH_3COOH + NaHCO_3 \rightarrow CH_3COO^-Na^+ + H_2O + CO_2$

  When a molecule **accepts** a proton H^+, then it is acting as a **base**.
  Example: $CH_3\mathbf{NH_2} + \mathbf{H}Cl \rightarrow CH_3\mathbf{NH_3^+}Cl^-$
- A molecule is **oxidised** when it **gains O** or **loses H**.
  Example: If you leave an unfinished bottle of wine opened overnight, then atmospheric oxygen oxidises the alcohol CH_3CH_2OH (ethanol) to vinegar CH_3COOH (ethanoic acid). The alcohol molecule has lost two H atoms and gained one O. In the laboratory, alcohols are oxidised by heating under reflux with an **oxidising agent**.

  A molecule is **reduced** when it **loses O** or **gains H**.
  Example: Ethanal CH_3CHO gains two H atoms when it is reduced to ethanol CH_3CH_2OH. A powerful **reducing agent** is needed. (See Fig. 8.2.)
- **Condensation reactions** occur when two or more molecules join and a small molecule is given off (often H_2O or HCl). Condensation is **addition** followed by **elimination**. **Example:** Ethanoic acid CH_3COOH and ethanol CH_3CH_2OH condense to form the ester $CH_3COOCH_2CH_3$ with the elimination of water H_2O. (See Fig. 8.3.)
- **Hydrolysis reactions** occur when **water** breaks a bond in a molecule. **Example:** Heating the **ester** ethyl ethanoate with water splits it into **ethanoic acid** and **ethanol**. An equilibrium mixture forms very slowly.

  $CH_3COOCH_2CH_3 + H_2O \rightleftharpoons CH_3COOH + CH_3CH_2OH$
- **Hydration reactions** involve the **addition** of **water** to a molecule.
  Example: $H_2O + CH_2CH_2 \rightarrow CH_3CH_2OH$ (See Fig. 8.4.)
- **Dehydration reactions** occur when **water** is **removed** from a molecule. **Example: Ethanol** is dehydrated to **ethene** when heated with the **dehydrating agent**. Each molecule loses 2 H atoms and one O atom.

  $CH_3CH_2OH \rightarrow CH_2CH_2 + H_2O$
- **Substitution reactions** occur when one group of atoms is replaced by another group. **Example:** $CH_3CH_2Br + OH^- \rightarrow CH_3CH_2OH + Br^-$

  Elimination reactions occur when some **atoms** are **removed** from an organic molecule. Remember that the elimination of **water** is called **dehydration**. You need to know about only one type of elimination reaction. **Example:** Heating **bromoethane** with KOH in pure ethanol forms **ethene** by the elimination of HBr.

  $CH_3CH_2Br + KOH \rightarrow CH_2CH_2 + H_2O + KBr$
- **Addition reactions** occur when a group of atoms are added to a molecule and no atoms are lost. **Example: Hydrogen bromide** adds to **ethene** to make **bromoethane**. $CH_2CH_2 + HBr \rightarrow CH_3CH_2Br$
- A **reaction mechanism** explains how a reaction happens. It shows how **electrons move** and which **bonds form** or **break** in each of the **steps** that make up the overall reaction. When you know the mechanism you generally realise which **reagents** are needed. Many reagents can be classed as nucleophiles or as electrophiles.

You will find studying organic chemistry easier if you know which type or class a reaction belongs to. Recognising the **reaction class** helps you choose the correct reagents.

CH_3COOH (ethanoic acid) is acting as an acid because it donates H^+ to the base OH^-. Sodium hydrogencarbonate and carbonates can also act as bases.

CH_3NH_2 (methylamine) is acting as a base because the amino $-NH_2$ group is accepting a proton H^+ from the aqueous hydrochloric acid HCl.

Acid: donates H^+
Base: accepts H^+

Fig. 8.1

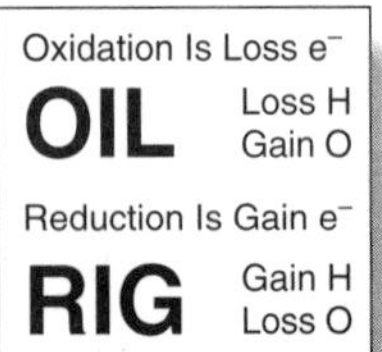

Oxidation Is Loss e⁻
OIL
Loss H
Gain O
Reduction Is Gain e⁻
RIG
Gain H
Loss O

Fig. 8.2

Condensation:
two molecules join and small molecule given off
Hydrolysis:
break bond + H_2O

Fig. 8.3

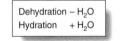

Dehydration $- H_2O$
Hydration $+ H_2O$

Fig. 8.4

- A **nucleophile** is a molecule or ion with an **electron-rich** site which can **donate** a pair of **electrons**. Nucleophiles **attack** the δ+ C atom of **weak bonds** in halogenoalkanes, aldehydes, or ketones. (See Fig. 8.5 for the nucleophilic substitution of bromomethane).

Nucleophiles include CN^-, OH^-, and Cl^-, as well as H_2O and NH_3. All nucleophiles have **lone pairs** which they **donate** to form **new bonds**.

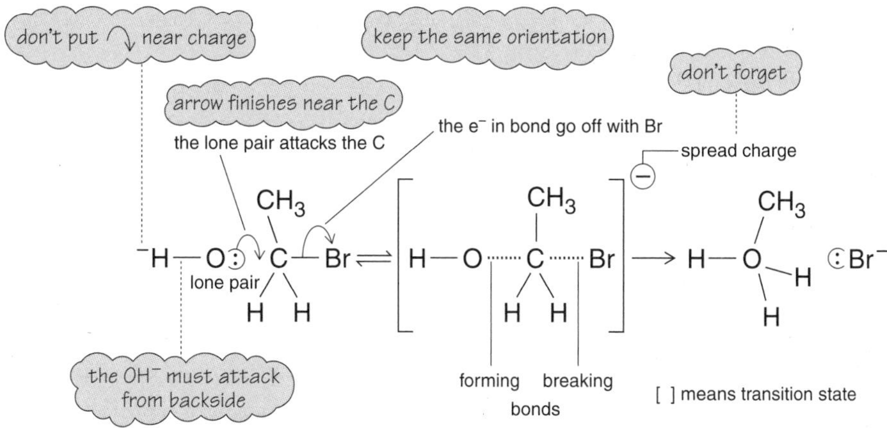

Fig. 8.5

In the substitution reaction given above, the **OH**⁻ ion is acting as a **nucleophile**. The reaction is therefore classed as a **nucleophilic substitution** reaction.

- An **electrophile** is a molecule or ion with an **electron-deficient** site which can **accept** a pair of **electrons**. Electrophiles react with the electron-rich pi bonds found in alkenes and aromatic molecules. (See Fig. 8.6 for the electrophilic addition of HBr to ethene.)

Electrophiles include positive ions such as H^+ and NO_2^+ and molecules such as Br_2 and HBr.

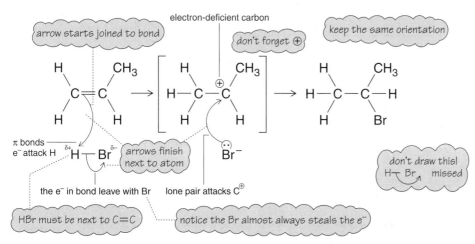

electron-deficient carbon

arrow starts joined to bond

don't forget ⊕

keep the same orientation

π bonds e⁻ attack H

arrows finish next to atom

don't draw this! missed

the e⁻ in bond leave with Br

lone pair attacks C⊕

HBr must be next to C=C

notice the Br almost always steals the e⁻

Fig. 8.6

In the addition reaction given above, the **HBr** is acting as an **electrophile**. The reaction is therefore classed as an **electrophilic addition** reaction.

- You will notice that when **bonds break**, the reaction mechanism usually shows that the **bonding pair** of electrons become located on **one atom**. This type of bond breaking is called **heterolytic fission**. Nucleophilic and electrophilic reactions usually involve heterolytic fission (see Fig. 8.7).

To show the **movement** of **two electrons** in a mechanism, you draw a **double-barbed curly arrow** (↷). Ensure you position the tail of the arrow on the lone pair, covalent bond, or pi bond and position the point exactly where the pair of electrons ends up.

Fig. 8.7

- A **free radical** is a molecule or atom with an **unpaired electron**.

- Free radical reactions involve the **movement** of **single electrons**. To show the movement of one electron, you draw a **single-barbed curly arrow** called a fish hook (⌒). When a bond breaks in a free radical reaction, the bonding electrons separate and one electron goes to each atom. This type of bond breaking is called **homolytic fission** (see Fig. 8.8).

A free radical is shown as a dot against the symbol or formula e.g. $Br^•$ or $CH_3^•$. An example of free radical substitution of alkanes is given in unit 9.

$Cl{-}Cl \rightarrow Cl^• + Cl^•$

Fig. 8.8

TESTS

RECALL TEST

1 Name the type of reagent required to carry out these changes:

a turn CH_3COOH into $CH_3COO^-Na^+$ ____________

b turn CH_3NH_2 into $CH_3NH_3^+$ ____________

c turn $C_6H_5CH_3$ into C_6H_5COOH ____________

d turn CH_3CN into $CH_3CH_2NH_2$ ____________

e turn CH_3CN into CH_3CONH_2 ____________

f turn CH_3CONH_2 into CH_3CN ____________ (6)

2 State the type of reaction:

a $CH_3CH_2OH + CH_3COCl \rightarrow CH_3COOCH_2CH_3 + HCl$

____________ (1)

b $CH_3CONHCH_2CH_3 + NaOH \rightarrow CH_3CO^-Na^+ + CH_3CH_2NH_2$

____________ (1)

3 What do these words mean?

a nucleophile

____

____________ (2)

b electrophile

____

____________ (2)

c addition

____

____________ (2)

d substitution

____

____________ (2)

4 What type of mechanism occurs in each of these reactions?

a $CH_3CH_2Cl + NaOH \rightarrow CH_3CH_2OH + NaCl$

____________ (2)

b $CH_3CHCH_2 + HBr \rightarrow CH_3CHBrCH_3$

____________ (2)

(Total 20 marks)

CONCEPT TEST

1 a Give an example of each of these reagent types:

Oxidising agent ______________________________

Reducing agent ______________________________

Dehydrating agent ______________________________ (3)

b Explain why converting ethanol into ethene is called dehydration.

______________________________ (1)

c Which reagent will convert ethanoic acid into sodium ethanoate?

______________________________ (1)

d How may sodium ethanoate be converted back to ethanoic acid?

______________________________ (1)

2 Ethanoic acid and ethanol react slowly to make ethyl ethanoate.

a What type of reaction does this illustrate?

______________________________ (1)

b Ethyl ethanoate may be split back into its components. Which chemical must be present for this to happen?

______________________________ (1)

3 a Explain what is meant by 'nucleophilic substitution'.

____

______________________________ (2)

b Explain what is meant by 'heterolytic fission'.

____

______________________________ (2)

c On a piece of paper, draw the mechanism of the reaction between bromoethane and potassium cyanide, KCN, which reacts in a similar way to sodium hydroxide, NaOH. (3)

4 Bromine, Br_2, reacts with ethene (containing >C=C<), but does not react with propanone (containing >C=O), which has a similar structure.

a Draw the mechanism for the reaction between ethene and bromine. (3)

b Why does >C=C< react with electrophiles and >C=O react with nucleophiles?

____

______________________________ (2)

(Total 20 marks)

Unit 9 ALKANES AND ALKENES

Alkanes have the general formula $\mathbf{C_nH_{2n+2}}$. They are called **saturated** hydrocarbons because all the C-C bonds are single bonds.

R, R', and R" all indicate an alkyl chain, e.g. a methyl group $-CH_3$.

- The alkanes are very **unreactive** (see unit 7). High-energy conditions are required to initiate reactions, that is **high temperature** (e.g. for combustion) or **ultraviolet light** (in free radical substitution reactions).

  Alkanes **combust** (burn) in air or oxygen. **Complete combustion** takes place in **excess oxygen** to form carbon dioxide and water. For example, in the case of methane: $CH_4 + 2O_2 \rightarrow CO_2 + 2H_2O$

- Alkanes react with halogens in ultraviolet light to make **halogenoalkanes**. An example is the bromination of methane.

  $CH_4 + Br_2 \rightarrow CH_3Br + HBr$

  Further substitution happens, producing CH_2Br_2, etc. The mechanism is by **free radical substitution**, which takes place in three steps: 1 **initiation**; 2 **propagation**; 3 **termination**.

  Initiation produces free radicals by the heterolytic fission of the Br-Br bond.

  $Br_2 \rightarrow 2Br^•$

  Propagation occurs when free radicals react with molecules to make other free radicals. Here, the $Br^•$ radical strikes the H atoms on the outside of the molecule (not the C atom):

  $CH_4 + Br^• \rightarrow CH_3^• + HBr$

  The product then forms, together with further free radicals.

  $CH_3^• + Br_2 \rightarrow CH_3Br + Br^•$

  Propagation is called a **chain reaction** because it uses up and then produces the reactive $Br^•$ radical.

  Termination happens when free radicals combine to produce neutral molecules, e.g.

  $2CH_3^• \rightarrow CH_3CH_3$

  $CH_3^• + Br^• \rightarrow CH_3Br$

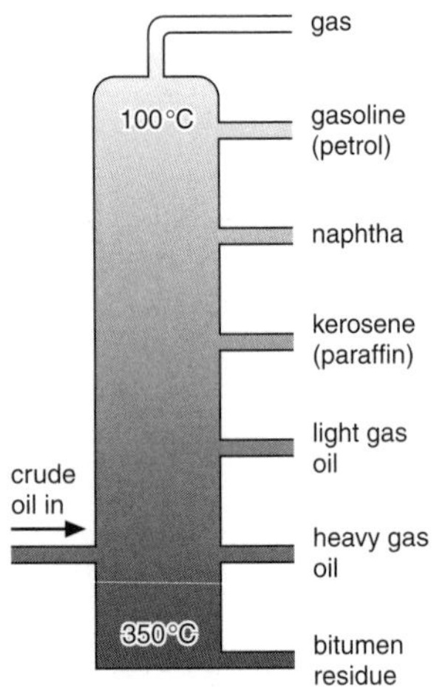

Fig. 9.1 The main fractions from the distillation of crude oil

The sources of alkanes are natural gas and crude oil (petroleum). **Natural gas** consists mostly of **methane**. **Crude oil** is a mixture of hundreds of different hydrocarbons, most of which are **alkanes**.

- **Fractional distillation** separates crude oil into groups of hydrocarbons called **fractions**. The temperature **decreases** up the fractionating column. Fractions containing low b.p. gases come from the top of the column; high b.p. substances that solidify at room temperature come from the bottom (see Fig. 9.1).

  Excessive amounts of the solid bitumen residue and other long-chain hydrocarbons are usually produced. **Cracking** breaks them into more useful molecules with shorter chains and also produces ethene.

  Thermal cracking uses high temperature and pressure to split long-chain alkanes into short-chain alkanes and alkenes (by a free radical mechanism). Hydrogen is a useful by-product.

  Catalytic cracking uses low pressure, high temperature, and zeolite catalysts to split long-chain alkanes into fractions used to make petrol, together with arenes ('aromatic' hydrocarbons containing benzene rings) (by a carbon cation (C^+) mechanism).

  Catalytic reforming is a similar process to catalytic cracking. The process also produces **cyclic** and **aromatic** compounds.

  Isomerisation is one form of reforming. It converts straight-chain alkanes into **branched-chain alkanes** which improves the octane number of petrol to stop its pre-ignition, which damages engines.

- You should know how to write or draw **molecular**, **structural**, **linear**, **display** or **graphical**, and **skeletal formulae** (see Fig. 9.2).

- **Alkenes** react by **electrophilic addition**, which is explained in unit 8.

  Alkenes have the general formula C_nH_{2n}. They are very **reactive** because they contain **unsaturated** C=C double bonds (C=O and C≡N etc. bonds are also unsaturated). Reactions usually happen at room temperature.

- You must recall these reactions of ethene. They all occur by electrophilic addition.

  $CH_2CH_2 + HBr \rightarrow CH_3CH_2Br$ (room temperature)

  $CH_2CH_2 + Br_2(CCl_4) \rightarrow CH_2BrCH_2Br$ (room temperature)

  $CH_2CH_2 + Br_2(aq) \rightarrow CH_2BrCH_2OH + HBr$ (room temperature)

  This last reaction acts as a **test** for alkenes, because the red-brown colour of Br_2 rapidly disappears.

  $CH_2CH_2 + H_2O \rightarrow CH_3CH_2OH$ using H_3PO_4 or H_2SO_4 and heat under reflux.

- During **electrophilic addition** the intermediate forms a C^+ ion, called a **carbocation**. If the C^+ is **tertiary** (e.g. $(CH_3)_3C^+$, which has three -CH_3 groups attached to it) then the C^+ ion is relatively stable and the reaction proceeds steadily. The -CH_3 groups donate electron density which spreads out and stabilises the + charge. The **secondary** C^+ ion (e.g. $(CH_3)_2HC^+$) is less stable, and the **primary** C^+ ion (e.g. $CH_3H_2C^+$) is the least stable.

- Alkenes can be **oxidised. Example:**

  $CH_2CH_2 + [O] + H_2O \rightarrow CH_2OHCH_2OH$ (ethane-1,2-diol)

  [O] is provided by dilute $KMnO_4$. Further oxidation of the OH groups occurs if warm or concentrated reagents are used. Ethane-1,2-diol is used in **antifreeze** and in **polyester** manufacture.

  Hydrogenation uses a **nickel catalyst** at 200 °C and high pressure to add hydrogen atoms across the alkene double bond.

  $CH_2CH_2 + H_2 \rightarrow CH_3CH_3$

  Hydrogenation changes liquid vegetable oils into solid **margarine**.

  Epoxyethane is made industrially by passing a mixture of air (or **oxygen**) and **ethene** over a **silver catalyst** (see Fig. 9.3). There is a danger of explosion because the ethene/oxygen mix is explosive. Epoxyethane is very reactive because the bond angle is 60°, which introduces bond strain caused by repulsion between displaced electrons. Epoxyethane is used to manufacture **epoxy resins**.

- **Addition polymers** are so useful because they are inert and **unreactive**. The main disadvantage of these plastic materials is they are almost completely non-biodegradable. Plastic refuse builds up in landfill sites; some plastics rot slowly to produce poisonous gases.

  One alternative is to **incinerate** waste plastic, using the heat to generate electricity. There is the danger of **poisonous combustion products** being released, e.g. burning PVC evolves HCl fumes which add to acid rain.

  Waste plastics can be **recycled**, but they must be **sorted** because mixing plastics produces a soft, almost useless, product.

- Alkenes are used to make **addition polymers**, e.g. polyethene is made from ethene using an initiator, high temperature, and high pressure.

  $nCH_2CH_2 \rightarrow [\text{-}CH_2CH_2\text{-}]_n$ (n is a large number 2000–35 000)

- Other monomers (substituted alkenes) make a wide range of addition polymers. (See Fig. 9.4.)

Monomer name	Monomer structure	Polymer repeating unit	Polymer name	Polymer uses
ethene	CH_2CH_2	$[\text{-}CH_2CH_2\text{-}]_n$	poly(ethene)	plastic bags
chloroethene (vinyl chloride)	CH_2CHCl	$[\text{-}CH_2CHCl\text{-}]_n$	poly(chloroethene) (PVC)	flooring, clothes, pipes
propene	CH_2CHCH_3	$[\text{-}CH_2CH\text{-}]_n$ with CH_3 on the CH	poly(propene)	plastic bottles
tetrafluoroethene	CF_2CF_2	$[\text{-}CF_2CF_2\text{-}]_n$	poly(tetrafluoroethene) (PTFE)	non-stick coating

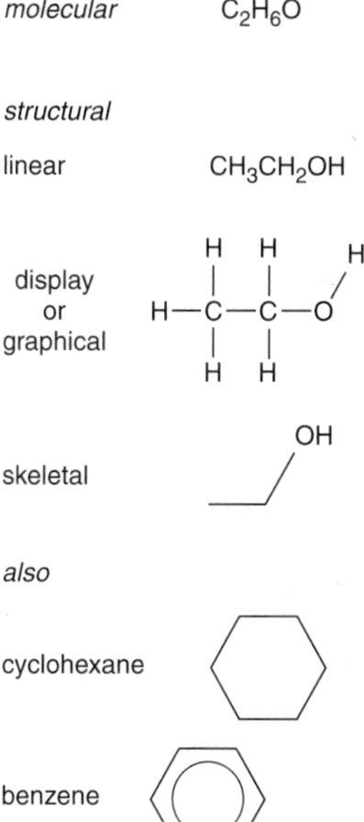

Fig. 9.2 Examples of different formula types

Fig. 9.3 Epoxyethane

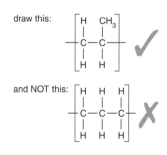

Fig. 9.4

Unit 9

TESTS

RECALL TEST

1 What is required to convert methane into chloromethane?

(2)

2 How can the chloromethane be converted into dichloromethane?

(2)

3 What is a free radical?

(2)

4 Write equations for the free radical substitution of ethane by bromine. You should label the stages initiation, propagation, and termination.

(6)

5 How are the components of crude oil separated?

(2)

6 Give two uses for the cracking of alkanes.

(2)

7 What mechanism is associated with the reactions of alkenes?

(1)

8 Generally what conditions are required when alkenes react?

(1)

9 Finish these equations:

a $CH_2CH_2 + HBr \rightarrow$

b $CH_2CH_2 + Br_2(CCl_4) \rightarrow$

c $CH_2CH_2 + Br_2(aq) \rightarrow$

d $CH_2CH_2 + H_2O \rightarrow$ (4)

10 Write a balanced equation for the action of dilute potassium permanganate (manganate(VII)) on ethene. You may use [O] to represent the manganate ion.

(1)

11 How could hexene be converted into hexane?

(3)

12 What is the test for alkenes?

(1)

13 Give the repeating unit for the polymer made when CH_3CHCH_2 polymerises.

(1)

14 What is the major disadvantage of polyalkanes?

(2)

(Total 30 marks)

CONCEPT TEST

1 Alkanes are a rich source of useful chemicals.

a Give the reagents and conditions necessary to make methane into tetrachloromethane.

______________________________ (2)

b In the laboratory, how could ethene be converted into ethanol?

______________________________ (2)

c Industrially, tonnes of ethene are made into ethane-1,2-diol. How could this reaction be carried out on a small scale?

______________________________ (2)

d How may octane be made from octene?

______________________________ (2)

2 a Some vegetable oils contain long-chain unsaturated molecules. Which chemical reaction would show that palm oil is saturated like animal fat, while sunflower oil is unsaturated? Give the reagent, conditions, and observations.

Reagent ______________________________

Conditions ______________________________

Observation with palm oil ______________________________

Observation with sunflower oil ______________________________ (4)

b Biodegradable polymers could be made from vegetable oil. Give one advantage and one disadvantage of biodegradable polymers.

____

______________________________ (2)

3 Free radicals are damaging to humans. The main sources are sunlight, smoke, and certain reactive chemicals.

a What is a free radical?

______________________________ (2)

b Explain how sunlight produces free radicals in your skin.

______________________________ (1)

c For a -CH_2- group on a molecule, show how it can be converted into -CHCl- by chlorine and sunlight.

____

______________________________ (3)

(Total 20 marks)

Unit 10

HALOGENOALKANES AND ALCOHOLS

Fig. 10.1 Bromoethane

- **Halogenoalkanes** (alkyl halides) consist of alkane molecules that have one or more hydrogen atoms replaced by **halogens** (F, Cl, Br, I). An example is bromoethane CH_3CH_2Br (see Fig. 10.1). Halogen atoms have greater numbers of electrons than hydrogen atoms, increasing the **induced Van der Waals forces**. The covalent C-halogen bond is polar (with the exception of C-I), producing **dipole–dipole interactions** between molecules. These intermolecular forces cause halogenoalkanes to have higher boiling points than the corresponding alkanes.

  Halogenoalkanes usually react by **nucleophilic substitution** by species such as OH^-, NH_3, and CN^- (see Fig. 10.2 and unit 8). Note that the **conditions** for these reactions include **heating under reflux** (see Fig. 10.3).

You will learn faster if you draw the mechanisms of these reactions.

Fig. 10.2 Nucleophilic substitution

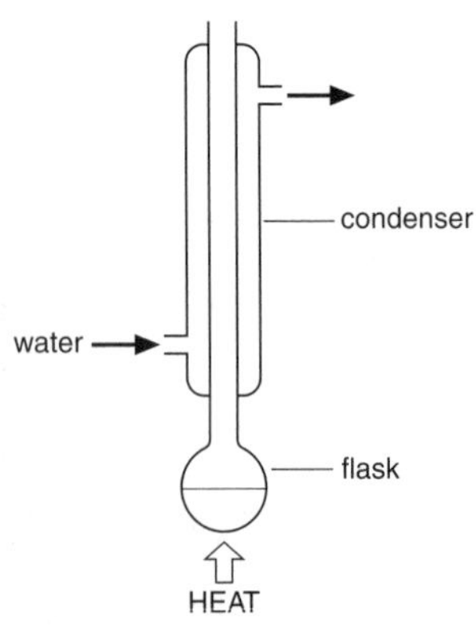

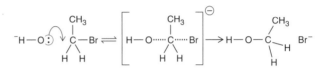

Fig. 10.3 Heat under reflux

- **Aqueous hydroxide** ions from NaOH act as a nucleophile and substitute the Br atom to form an **alcohol**. This reaction is often referred to as **alkaline hydrolysis**.

  $CH_3CH_2Br + KOH(aq) \rightarrow CH_3CH_2OH + KBr$

- **Alcoholic hydroxide** ions from KOH dissolved in dry ethanol convert a halogenoalkane into an **alkene** by **elimination**. (See Fig. 10.4.)

  $CH_3CH_2Br + KOH(alcohol) \rightarrow CH_2CH_2 + KBr + H_2O$

  The OH^- ions are unable to act as nucleophiles because they are attached by hydrogen bonding to the alcohol molecules. They can still act as a base, accepting protons H^+. Removal of H^+ from bromoethane causes a Br^- ion to also leave; the electron pair from the broken C-H bond then forms an alkene pi bond.

- **Alcoholic ammonia** forms **amines** when heated with halogenoalkanes.

  $CH_3CH_2Br + 2NH_3(alcohol) \rightarrow CH_3CH_2NH_2 + NH_4Br$

  The reaction produces a low yield because the (**primary**) amine is also a nucleophile and will attack bromoethane to produce **secondary**, **tertiary**, and **quaternary** amines.

  $CH_3CH_2NH_2 \rightarrow (CH_3CH_2)_2NH \rightarrow (CH_3CH_2)_3N \rightarrow (CH_3CH_2)_4N^+$

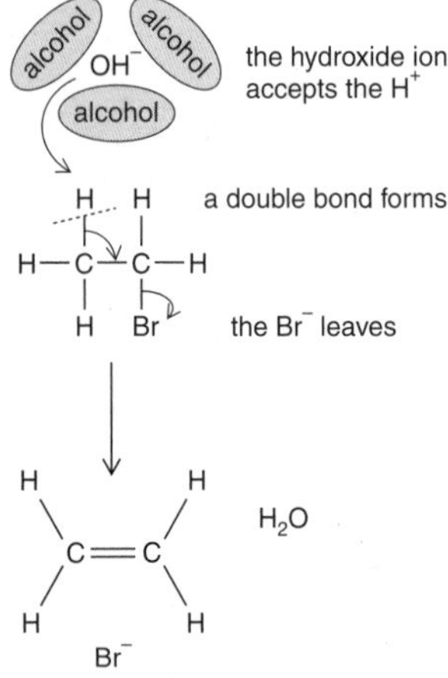

Fig. 10.4

- **Alcoholic cyanide** ions form **nitriles**.

  $CH_3CH_2Br + KCN(alcohol) \rightarrow CH_3CH_2CN + KBr$

  This reaction **increases** the carbon chain by **one C atom**. A nitrile made from a halogenoalkane can be turned into a **carboxylic acid** by heating under reflux in aqueous acid. The reaction is often referred to as **acid hydrolysis**, e.g.

  $CH_3CH_2Br \rightarrow CH_3CH_2CN \rightarrow CH_3CH_2COOH$

  You do not have to memorise a balanced equation.

- The **test** for a halogenoalkane is to add aqueous acidified **silver nitrate**. The water acts as a nucleophile and slowly substitutes OH for the halogen.

  $R\text{-}Hal + H_2O \rightarrow R\text{-}OH + HHal$

  The halide ion released then combines with the silver ions to form a **precipitate**.

  $Ag^+ + Hal^- \rightarrow AgHal$

  The precipitate colours are: iodoalkane → **pale yellow** (AgI); bromoalkane → **off white** (AgBr); chloroalkane → **white** (AgCl).

- The reactivity of halogenoalkanes make them useful **intermediates in industry**. (See unit 16.)

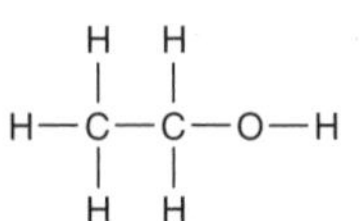

Fig. 10.5 Ethanol

- **Alcohols** consist of a **hydroxyl group** -OH covalently bonded to a hydrocarbon. Ethanol CH_3CH_2OH (see Fig. 10.5) is an example of an aliphatic alcohol, in which -OH groups are bonded to straight or branched chain hydrocarbons. The O-H bond is **polar** $O^{\delta-}$—$H^{\delta+}$, which causes **hydrogen bonding**. Alcohols have much higher boiling points than the corresponding alkanes e.g. ethanol (M_r 46) 76 °C; propane (M_r 44) b.p. –42 °C). The H-bonds also enable alcohols to **dissolve** in water (solubility decreases as the non-polar hydrocarbon chain increases in size).
- You should know the difference between **primary**, **secondary**, and **tertiary** alcohols (see Fig. 10.6).
- **Halogenation** involves nucleophilic substitution to convert alcohols into the corresponding halogeno compound, e.g. ethanol to the **halogenoalkanes** chloroethane, bromoethane, and iodoethane:

  $CH_3CH_2OH + PCl_5 \rightarrow CH_3CH_2Cl + POCl_3 + HCl$ (room temperature)

  $CH_3CH_2OH + HBr \rightarrow CH_3CH_2Br + H_2O$
  (heat under reflux, with HBr made in situ using NaBr and conc. H_2SO_4)

  $CH_3CH_2OH + PI_3 \rightarrow CH_3CH_2I + PIO + HI$
  (room temperature with PI_3 made in situ by mixing iodine with red phosphorus)
- Alcohols may be **oxidised** by combustion. Ethanol burns with a clean blue flame. $CH_3CH_2OH + 3O_2 \rightarrow 2CO_2 + 3H_2O$

  When exposed to the air, **primary alcohols** RCH_2OH will **oxidise** very slowly to **aldehydes** RCHO and then to carboxylic acids RCOOH (e.g. beer and wine slowly change to vinegar). The strong oxidising agents acidified potassium dichromate(VI) $K_2Cr_2O_7$ and acidified potassium permanganate manganate(VII) $KMnO_4$ are used in the lab.
- To stop the oxidation at the **aldehyde**, the oxidising agent $K_2Cr_2O_7$ with H_2SO_4 is dripped into hot **primary alcohol**, and the aldehyde is **distilled** off as it forms.

  $CH_3CH_2OH + [O] \rightarrow CH_3CHO + H_2O$

  Notice you just write [O] to show there is an oxidising agent.

  The aldehyde has a polar C=O so has a lower b.p. (49 °C) than the hydrogen-bonded alcohol (76 °C). Use $NaBH_4(aq)$ to reduce back to the alcohol.

  $CH_3CHO + 2[H] \rightarrow CH_3CH_2OH$

  To oxidise a primary alcohol to the carboxylic acid, heat under reflux with the oxidising agent and then distil off the product.

  $CH_3CH_2OH + 2[O] \rightarrow CH_3COOH + H_2O$
  (Use $LiAlH_4$ to reduce back to the alcohol.)
- **Secondary alcohols** oxidise to form a **ketone** e.g. propan-2-ol plus oxidising agent form propanone and water.

  $CH_3CH(OH)CH_3 + 2[O] \rightarrow CH_3COCH_3 + H_2O$
- **Tertiary alcohols** cannot easily be oxidised except by combustion.
- All alcohols can be **dehydrated** to make **alkenes** (see Fig. 10.7). The reagents and conditions are to either heat under reflux with conc. H_2SO_4 or H_3PO_4 or to pass the alcohol vapour over hot pumice or Al_2O_3.
- All alcohols will join with carboxylic acids in condensation reactions to form **esters**. $CH_3CH_2OH + CH_3COOH \rightarrow CH_3COOCH_2CH_3 + H_2O$

  The reaction happens when the two compounds are mixed and warmed. Adding concentrated H_2SO_4 catalyses the reaction (increases the rate) and removes the water (increases the yield).
- Hydrogen gas is steadily evolved when **sodium metal** reduces ethanol to make sodium **ethoxide** (a substance that is useful in organic synthesis).

  $2CH_3CH_2OH + 2Na \rightarrow 2CH_3CH_2O^-Na^+ + H_2$

You need to recall the **five reaction types** of alcohols – **halogenation**, **oxidation**, **dehydration**, **esterification**, and **reduction**.

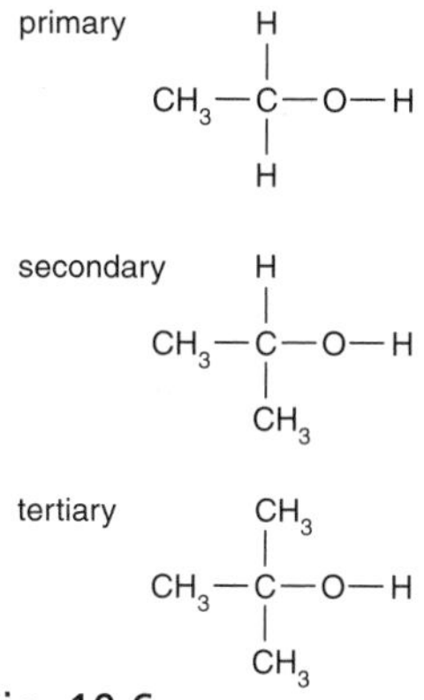

primary: $CH_3—CH_2—O—H$

secondary: $CH_3—CH(CH_3)—O—H$

tertiary: $CH_3—C(CH_3)_2—O—H$

Fig. 10.6

For most reactions you write 'heat under reflux' for the conditions (see Fig 8.3), but some reactions do occur at room temperature.

Converting alcohols into halogenoalkanes is useful in organic synthesis because you can convert them into many other types of compound.

To **test** for an alcohol, suggest adding $\mathbf{PCl_5}$. Dense white fumes (of HCl) show an OH group is present. NB also gives white fumes with water and carboxylic acids.

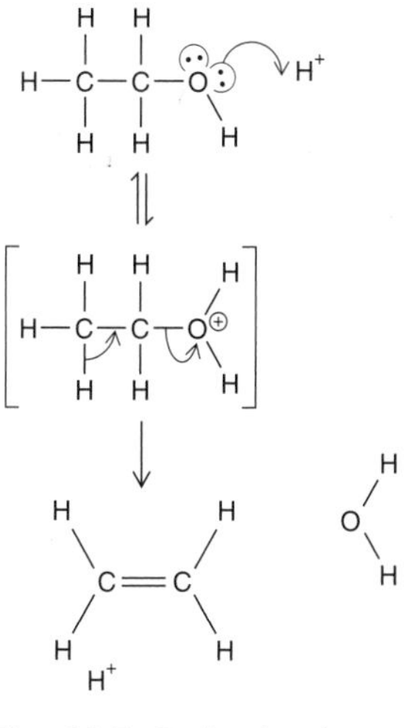

H^+

Fig. 10.7 Dehydration of alcohol

TESTS

RECALL TEST

1 Name the mechanism which predominates in the halogenoalkane reactions.

______________________________ (2)

2 Finish these equations.

a $CH_3CH_2Cl(l) + NaOH(aq) \rightarrow$ ______________

b $CH_3CH_2I(l) + NaOH(ethanol) \rightarrow$ ______________

c $CH_3CH_2Br(l) + KCN(ethanol) \rightarrow$ ______________

d $CH_3CH_2Br(l) + KOH(ethanol) \rightarrow$ ______________ (8)

3 Identify the product formed when CH_3CH_2CN is boiled under reflux with aqueous hydrochloric acid.

______________________________ (2)

4 State the reagents and conditions required to convert CH_3CN into CH_3COOH.

______________________________ (2)

5 Hydroxide ions react with bromoethane in two different ways, depending on the conditions. State the two conditions and the two organic products.

____

______________________________ (4)

6 Aqueous silver nitrate is mixed with an organic compound to produce a cream precipitate. Identify the cream precipitate and the functional group in the original organic compound.

______________________________ (2)

7 Why are short-chain alcohols soluble in water, while long-chain alcohols are insoluble?

______________________________ (2)

8 Why does ethanol have a much higher boiling point than ethanal?

______________________________ (2)

9 Name these alcohols: **a** CH_3CH_2OH ______________

b $CH_3CHOHCH_3$ ______________ **c** $(CH_3)_3COH$ ______________

d $(CH_3)_3CCH_2OH$ ______________ (8)

10 Fill in the rest of these equations.

a $CH_3CH_2OH + Na \rightarrow$ ______________

b $CH_3CH_2OH + [O] \rightarrow$ ______________ using $KMnO_4(aq)$ with $H_2SO_4(aq)$

c $CH_3CH_2OH \rightarrow$ ______________ using concentrated $H_2SO_4(l)$

d $CH_3COOH + CH_3CH_2OH \rightarrow$ ______________ with concentrated $H_2SO_4(l)$

e $CH_3CH_2OH + PCl_5 \rightarrow$ ______________

f $CH_3CH_2OH + HBr \rightarrow$ ______________

g $CH_3CHOHCH_3 + [O] \rightarrow$ ______________ (8)

(Total 40 marks)

CONCEPT TEST

1 Give examples of bromo compounds with a molecular formula C_4H_9Br:

a a primary halogenoalkane with a branching side chain,

____ (1)

b a secondary halogenoalkane.

____ (1)

2 a 2-bromopropane will react in two ways with KOH depending on the conditions. State the two conditions and name the organic products.

Condition 1 ______________ Product 1 ______________ (2)

Condition 2 ______________ Product 2 ______________ (2)

b Both **i** 2-aminopropane and **ii** 2-methylpropanenitrile may be formed from 2-bromopropane. Give the reagents and conditions.

i Formation of 2-aminopropane:

Reagents ______________ Conditions ______________ (2)

ii Formation of 2-methylpropanenitrile:

Reagents ______________ Conditions ______________ (2)

c How could you show in the laboratory that a compound contained a C-Br group?

____ (2)

3 Give the structural formula of an isomer of $C_4H_{10}O$ which is:

a a primary alcohol ______________________________

b a secondary alcohol ______________________________

c a tertiary alcohol ______________________________ (3)

4 Here are three reactions of propan-2-ol:

$$CH_3CHOHCH_3$$

$$\swarrow A \qquad \downarrow B \qquad \searrow C$$

$$CHCOOCH(CH_3)_2 \qquad CH_3CHBrCH_3 \qquad CH_3CH{=}CH_2$$

a Give the reagents and conditions to convert propan-2-ol for the reactions A to C:

Reaction A: Reagents ______________ Conditions ______________

Reaction B: Reagents ______________ Conditions ______________

Reaction C: Reagents ______________ Conditions ______________ (6)

b Propan-2-ol will also form an ester with ethanoic acid. Give the structural formula of this ester.

____ (2)

c 2-methylpropan-2-ol reacts differently to propan-2-ol. Identify the products when 2-methylpropan-2-ol reacts with the following. If the reagent does not react with 2-methylpropan-2-ol, then state that it does not react.

i concentrated sulphuric acid ______________________________

ii potassium dichromate ______________________________ (2)

(Total 25 marks)

Unit 11

ENERGETICS: ENTHALPY CHANGE

You must remember that an enthalpy change is a **heat change** and **not** an energy change (which can involve doing work).

- The **heat** evolved from or absorbed by a reaction at constant pressure is called the **enthalpy change**.

  When a reaction gives out heat, we say that the reaction is **exothermic**. The heat change (units $J\,mol^{-1}$) is given a **negative sign** because the reacting chemicals have **lost heat** to their surroundings. Combustion of a piece of paper and respiration are examples of exothermic reactions.

  When a reaction takes in heat, we say that the reaction is **endothermic**. The heat change is given a **positive** sign because the reacting chemicals have **gained heat** from their surroundings. An example is photosynthesis.

Enthalpy changes also occur during **physical changes** such as boiling and freezing.

- Enthalpy changes are measured for **1 mole** of substance under standard conditions (**298 K**, 1 atmosphere pressure, **101 325 Pa**). When you define a standard enthalpy change remember to state the standard conditions. **Example:** The standard enthalpy of vaporisation of water:

  $H_2O(l) \rightarrow H_2O(g) \quad \Delta H_v^\ominus = +41.1\ kJ\,mol^{-1}$

- You must recall the definitions (but not numerical values) of certain standard enthalpy changes.

  Enthalpy of formation: The enthalpy change when one mole of a substance is formed from its constituent elements in their normal states (under standard conditions). **Example:**

  $Na(s) + \frac{1}{2}Cl_2(g) \rightarrow NaCl(s) \quad \Delta H_f^\ominus(NaCl) = -411\ kJ\,mol^{-1}$

  Enthalpy of combustion: The enthalpy change when one mole of a substance is completely combusted in oxygen (under standard conditions). **Example:**

  $CH_4(g) + 2O_2(g) \rightarrow CO_2(g) + 2H_2O(l) \quad \Delta H_c^\ominus(\text{methane}) = -890.4\ kJ\,mol^{-1}$

  Enthalpy of neutralisation: The enthalpy change when one mole of water is formed from the reaction of an acid and a base (under standard conditions). **Example:**

  $NaOH(aq) + HCl(aq) \rightarrow NaCl(aq) + H_2O(l) \quad \Delta H_n^\ominus = -57.1\ kJ\,mol^{-1}$

  Enthalpy of reaction: The enthalpy change that accompanies a reaction between the amounts of substances (under standard conditions) shown in the balanced chemical equation. **Example:** The reaction between ammonia and fluorine.

  $NH_3(g) + 3F_2(g) \rightarrow 3HF(g) + NF_3(g) \quad \Delta H_r^\ominus = -875\ kJ\,mol^{-1}$

  Mean bond enthalpy: The **average** energy required to **break** one mole of a particular kind of bond derived from a wide range of molecules that contain the bond. The environment of a given bond type may be different in different molecules. As a result, mean bond enthalpy values will not exactly agree with bond enthalpy values derived from one particular molecule. **Example:**

  Mean bond enthalpy$_{(C\text{-}H)}$ = $+412\ kJ\,mol^{-1}$

The enthalpy change for **making** a bond has the same value but opposite sign.

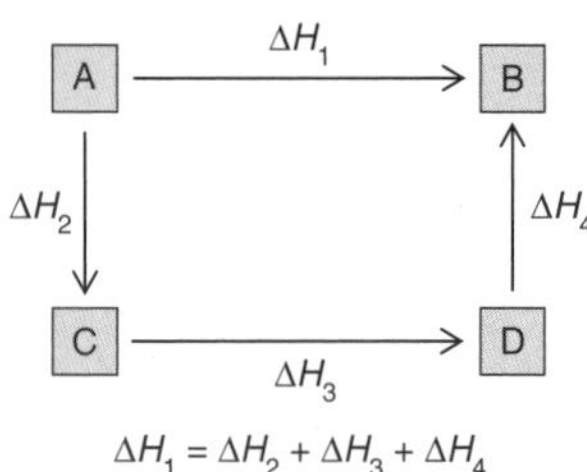

$\Delta H_1 = \Delta H_2 + \Delta H_3 + \Delta H_4$

Fig. 11.1 Hess's law

- **Hess's law** states that the enthalpy change accompanying a reaction is **independent of the route** taken. Suppose A → B directly and also A → C → D → B indirectly. The enthalpy change accompanying A → B equals the sum of the enthalpy changes accompanying A → C, C → D, and D → B (see Fig. 11.1).

- There are many ways of organising the Hess's law **calculations** you will meet at AS level. Here is a good way to do calculations that will produce fewer mistakes (see Figs 11.2, 11.3, and 11.4).

  One: Translate all the ΔH terms given in the question into chemical equations.

  Two: Across the full width of the page, write out the chemical equation that corresponds to the enthalpy change you seek. This equation represents the **direct route**.

If you have found a way that works for you AND you get correct answers most of the time, then stick with it.

Remember that you must know the definitions given in this spread: you cannot start to attempt calculations without them.

Three: There will be an **indirect route** between the reactants and the products written in step **two**. Inspect the data you are given to see if there are obvious intermediate substances. If you are given ΔH_f, then **intermediates** might be elements; if ΔH_c (for hydrocarbons), then intermediates could be CO_2 and H_2O.

Four: Set out the Hess's law **enthalpy** cycle. Make sure the **arrows** point in the **correct directions**.

Example: Direct route: elements → combustion products; indirect route: elements → compound → combustion products. **Remember** to (i) write the values for enthalpy changes over the arrows; (ii) reverse the signs of the enthalpy changes for reactions that reverse the given data; (iii) multiply molar enthalpy changes according to the number of moles of substance in the equations.

Example: If two H_2O molecules are made from the combustion of hydrogen, then you must write the total enthalpy change as $2 \times \Delta H^\ominus_{c(H_2)}$.

Five: Draw an arrow from the starting substances to the end products via the intermediate reaction(s). **Sum** the enthalpy changes for the indirect route and **equate** these to the enthalpy change for the direct route.

Check that you have **balanced** the equations and used them correctly to multiply the values of the ΔH terms. Check you have drawn the arrows in the **correct direction** and that the signs of the enthalpy change terms are appropriate to the direction of the change concerned.

Sometimes the **examiners** will give you data in a **previous part** of the question; sometimes they will give you **surplus** data.

To work out the heat change, use the relationship

$q = mc\Delta T$

q = quantity of heat
m = mass
c = specific heat capacity
ΔT = temperature change

Usually the mass concerned is for water (density = 1 g cm^{-3}; c = 4.2 J g^{-1} K^{-1}).

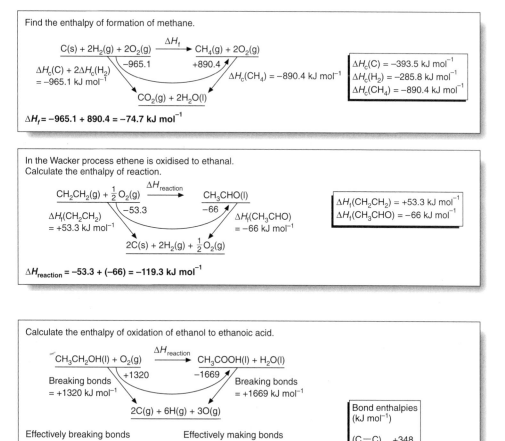

Find the enthalpy of formation of methane.

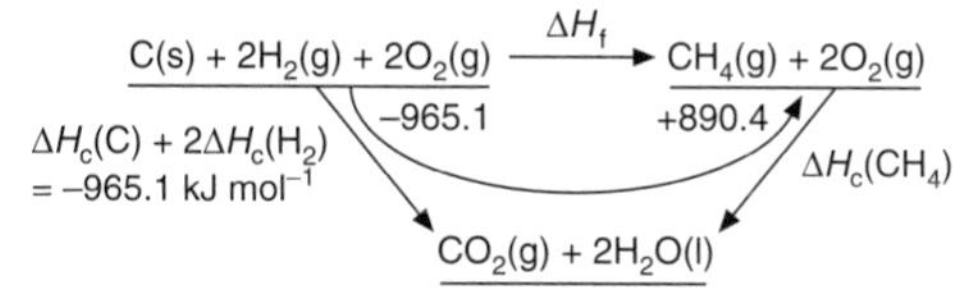

$C(s) + 2H_2(g) + 2O_2(g) \xrightarrow{\Delta H_f} CH_4(g) + 2O_2(g)$

$\Delta H_c(C) + 2\Delta H_c(H_2) = -965.1$ kJ mol^{-1} ↓ −965.1 → $CO_2(g) + 2H_2O(l)$ ← +890.4; $\Delta H_c(CH_4) = -890.4$ kJ mol^{-1}

$\Delta H_c(C) = -393.5$ kJ mol^{-1}
$\Delta H_c(H_2) = -285.8$ kJ mol^{-1}
$\Delta H_c(CH_4) = -890.4$ kJ mol^{-1}

$\Delta H_f = -965.1 + 890.4 = -74.7$ kJ mol^{-1}

Fig. 11.2

In the Wacker process ethene is oxidised to ethanal.
Calculate the enthalpy of reaction.

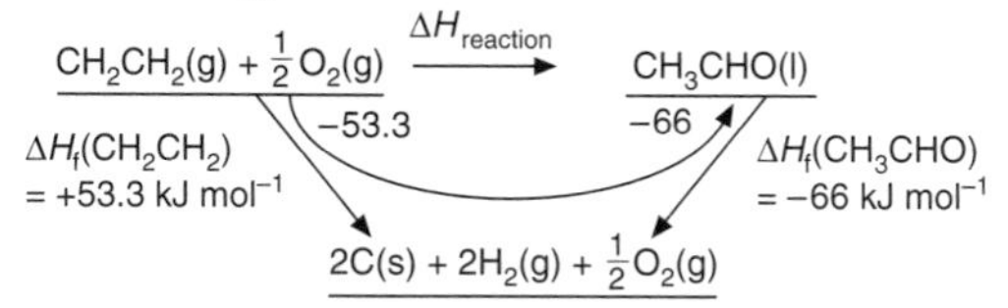

$CH_2CH_2(g) + \frac{1}{2}O_2(g) \xrightarrow{\Delta H_{reaction}} CH_3CHO(l)$

$\Delta H_f(CH_2CH_2) = +53.3$ kJ mol^{-1} ↓ −53.3 → $2C(s) + 2H_2(g) + \frac{1}{2}O_2(g)$ ← −66; $\Delta H_f(CH_3CHO) = -66$ kJ mol^{-1}

$\Delta H_f(CH_2CH_2) = +53.3$ kJ mol^{-1}
$\Delta H_f(CH_3CHO) = -66$ kJ mol^{-1}

$\Delta H_{reaction} = -53.3 + (-66) = -119.3$ kJ mol^{-1}

Fig. 11.3

Calculate the enthalpy of oxidation of ethanol to ethanoic acid.

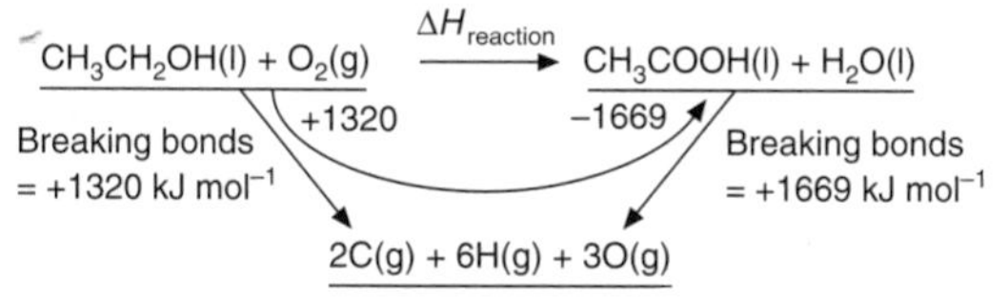

$CH_3CH_2OH(l) + O_2(g) \xrightarrow{\Delta H_{reaction}} CH_3COOH(l) + H_2O(l)$

Breaking bonds = +1320 kJ mol^{-1} ↓ +1320 → $2C(g) + 6H(g) + 3O(g)$ ← −1669; Breaking bonds = +1669 kJ mol^{-1}

Effectively breaking bonds		Effectively making bonds	
2 × (C—H) = 2 × (+412) =	+824	1 × (C=O)	−743
1 × (O=O) =	+496	2 × (O—H) = −2 × (+463) =	−926
Total =	+1320	Total =	−1669
(Breaking bonds is endothermic)		(Making bonds is exothermic)	

Bond enthalpies (kJ mol^{-1})	
(C—C)	+348
(C—H)	+412
(C—O)	+360
(C=O)	+743
(O—H)	+463
(O=O)	+496

Enthalpy of reaction = +1320 − 1669 = −349 kJ mol^{-1}

Fig. 11.4

ΔH_f: elements → compound

ΔH_c: compound + O_2 → combustion products (often CO_2, H_2O)

$\Delta H_{neutralisation}$: acid + base → water (+ a salt)

ΔH_{bond}: gaseous molecule → gaseous atoms

$\Delta H_{solution}$: solid + (aq) → aqueous ions

1st ionisation energy: gaseous atom → gaseous (+) ion

1st electron affinity: gaseous atom → (–) gaseous ion

Fig. 11.5 The enthalpies in short

TESTS

RECALL TEST

1 What sign does an enthalpy have in an exothermic reaction?

(1)

2 Define the enthalpy of:

a formation

(3)

b combustion

(3)

c neutralisation

(3)

3 Define 'bond enthalpy'.

(3)

4 Define 'mean bond enthalpy'.

(3)

5 State Hess's law.

(2)

6 Write the equation used to calculate the heat involved in changing the temperature of a mass of water.

(2)

(Total 20 marks)

CONCEPT TEST

1 As crude oil is going to run out, research has focussed on making organic compounds from coal. One idea is to react coal with water to make carbon monoxide which then reacts with hydrogen to make methanol.

$\Delta H_f(CO_2(g)) = -394\ kJ\ mol^{-1}$

$\Delta H_f(CO) = -111\ kJ\ mol^{-1}$

$\Delta H_f(H_2O(l)) = -286\ kJ\ mol^{-1}$

$\Delta H_c(CO) = -283.0\ kJ\ mol^{-1}$

$\Delta H_c(CH_3OH) = -715.0\ kJ\ mol^{-1}$

Reaction A: $C(s) + H_2O(l) \rightarrow CO(g) + H_2(g)$

Some enthalpies are listed in the margin.

a Use the data above to calculate the enthalpy of the reaction A.

____

____ (4)

b An alternative idea is to convert coal and water in the presence of hydrogen directly to methanol (but a catalyst has yet to be perfected).

Reaction B: $C(s) + H_2O(l) + H_2(g) \rightarrow CH_3OH(l)$

Calculate the enthalpy of reaction B.

____

____ (4)

c The methanol could then be converted to methanal, CH_2O.

Here are some bond enthalpies, in $kJ\ mol^{-1}$.

C-C	C-H	C-O	C=O	O-H	O=O
348	412	360	743	463	496

Calculate the enthalpy of reaction for the oxidation of methanol:

$CH_3OH(l) + \frac{1}{2}O_2(g) \rightarrow CH_2O(g) + H_2O(l)$

____

____ (3)

2 Given these enthalpies, calculate the enthalpy of formation of dinitrogen tetraoxide, N_2O_4:

$2NO_2(g) \rightarrow N_2O_4(g)$
$\Delta H_{reaction} = -58.1\ kJ\ mol^{-1}$

$\frac{1}{2}N_2(g) \rightarrow \frac{1}{2}O_2(g) \rightarrow NO(g)$
$\Delta H_f(NO) = +90.4\ kJ\ mol^{-1}$

$NO(g) + \frac{1}{2}O_2(g) \rightarrow NO_2(g)$
$\Delta H_{reaction} = -56.5\ kJ\ mol^{-1}$

____

____ (2)

3 The oxidation of ammonia to make nitrogen monoxide is very important as the nitrogen monoxide may then be converted easily into nitric acid. The stoichiometric equation could be written as

$$4NH_3(g) + 5O_2(g) \rightarrow 4NO(g) + 6H_2O(l)$$

Various enthalpies of formation are shown right.

$\Delta H_f(NH_3) = -46.2\ kJ\ mol^{-1}$

$\Delta H_f(NO) = +90.4\ kJ\ mol^{-1}$

$\Delta H_f(H_2O) = -286\ kJ\ mol^{-1}$

a Use Hess's law to calculate the enthalpy of reaction for the oxidation of ammonia.

____

____ (3)

b Calculate the (Si-Cl) bond enthalpy in $SiCl_4$, given these values:

$\Delta H_f(SiCl_4(l)) = -640\ kJ\ mol^{-1}$; $\Delta H_a(Si) = +439\ kJ\ mol^{-1}$; (Cl-Cl) bond enthalpy = $+242\ kJ\ mol^{-1}$.

____

____ (4)

(Total 20 marks)

Unit 12

LATTICE ENTHALPY AND ENTROPY

- **Lattice enthalpy** is the standard enthalpy when a solid ionic lattice is broken into separate gaseous ions.
  Example: $CaCl_2(s) \rightarrow Ca^{2+}(g) + 2Cl^-(g) \quad \Delta H^{\ominus}_{latt} = +2237\ \text{kJ mol}^{-1}$
- **Lattice formation entl** [...] ndard enthalpy change when a solid ionic lattice is **formed** f [...] ions. Lattice formation enthalpy and lattice (di [...] e the same value, but opposite signs. Lattice [...] ys has a negative sign.
- Ions with **higher cha** [...] **ller ionic radius**) produce a **greater** lattice format [...] es of lattice enthalpy indicate **high melting points** [...]
  Example: Ions in M [...] ubly charged, so are strongly attracted to each oth [...] their lattice. MgO has a very high melting point ([...] ractory lining for furnaces.
  Enthalpy of atomis [...] ange when one mole of gaseous atoms is formed from an element in its [...] nal state under standard conditions. **Example:**
  $\frac{1}{2}H_2(g) \rightarrow H(g) \quad \Delta H^{\ominus}_{a}(H) = +218\ \text{kJ mol}^{-1}$
  Enthalpy of hydration: The enthalpy change when one mole of separate gaseous ions form hydrated ions (under standard conditions). **Example:**
  $Na^+(g) + \text{water} \rightarrow Na^+(aq) \quad \Delta H^{\ominus}_{hyd} = -406\ \text{kJ mol}^{-1}$
  Enthalpy of solution: The enthalpy change when one mole of a solid dissolves in a large excess of water (under standard conditions) to produce separate hydrated ions. **Example:**
  $NaOH(s) + \text{water} \rightarrow Na^+(aq) + OH^-(aq) \quad \Delta H^{\ominus}_{solu} = -42.7\ \text{kJ mol}^{-1}$

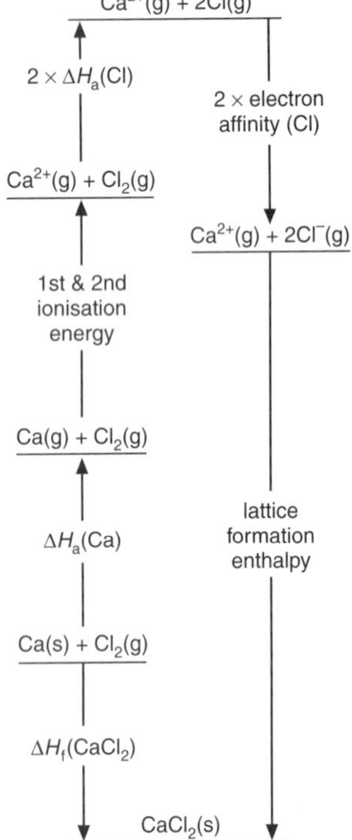

Fig. 12.1

- You must know how to set out **Born–Haber cycles** (diagrams) that use Hess's law to calculate lattice enthalpies. You will find it useful to refer to the Born–Haber diagram for NaCl found in many text books.
  Example: The Born–Haber diagram to calculate the lattice formation enthalpy of $CaCl_2$ (see Fig. 12.1).

  1st ionisation energy of Ca = +590 — $\Delta H_a(Cl) = +121$
  2nd ionisation energy of Ca = +1740 — $\Delta H_a(Ca) = +193$ — (data in kJ mol^{-1})
  electron affinity of Cl = –364 — $\Delta H_f(CaCl_2) = -795$

  Comparing the two sides, from top of diagram to bottom:
  $^{-}(2 \times {}^{+}121) + {}^{-}({}^{+}590 + 1150) + {}^{-}193 + {}^{-}795 = (2 \times {}^{-}728) +$ lattice energy
  Note the change in sign due to the change in enthalpy change direction.
  So the lattice formation enthalpy = –2242 kJ mol^{-1}.
  Learn to combine lattice formation enthalpy, hydration enthalpy, and solution enthalpy into a calculation and into one diagram (see Fig. 12.2).

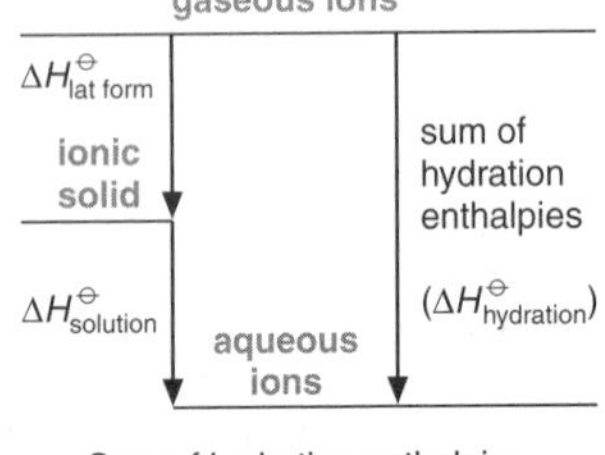

Sum of hydration enthalpies $= \Delta H^{\ominus}_{lat\ form} + \Delta H^{\ominus}_{solution}$

Fig. 12.2

- Lattice enthalpies **calculated** from Born–Haber cycles **differ** from those **predicted** by theoretical calculations based on a purely ionic model. See Fig. 12.3 for an example. These differences are evidence for the **covalent character** of many ionic solids.

	Experimental	Theoretical
AgF	960	920
AgCl	905	835
AgBr	890	815
AgI	890	780

Fig. 12.3 The lattice enthalpies (kJ mol^{-1}) of the silver halides

- **Thermal stabilities** may be explained in terms of the **polarisation** of the carbonate anion (see unit 3). There is also a relationship between thermal stabilities and the **lattice enthalpies** of carbonates and oxides. The oxide lattice enthalpies decrease down group 2 ($MgCO_3$ to $BaCO_3$) as the cation radius increases, so the carbonate lattices require successively higher temperatures to break down.

The trends in **solubilites** of group 2 hydroxides and sulphates were discussed in unit 3.

- Heat is evolved when solutions of an **acid** and an **alkali** are mixed. When dilute strong acid (e.g. aqueous HCl) neutralises dilute strong alkali (e.g. aqueous NaOH), the value of the **enthalpy of neutralisation** is always about –57 kJ mol^{-1}. Strong acids and strong alkalis are fully ionized in dilute solution, so the enthalpy change is always due to the reaction
  $H^+(aq) + OH^-(aq) \rightarrow H_2O(l)$
  All other ions are **spectator ions** and do not take part in the reaction.

- **Less heat** is evolved during neutralisation when either or both of the acid and alkali are **weak**.
  Example: The reaction between dilute sodium hydroxide and ethanoic acid.
  $CH_3COOH(aq) + NaOH(aq) \rightarrow CH_3CO_2^-Na^+(aq) + H_2O(l)$ $\Delta H^{\ominus}_{neut} = -55.2$ kJ mol^{-1}
  During the reaction, energy is required to dissociate the weak acid into ions.
  $CH_3COOH(aq) \rightarrow CH_3CO_2^-(aq) + H^+(aq)$
- **Entropy** is a measure of **disorder**. Entropy (symbol S) increases when matter or energy spread out. Everyday examples of entropy increase include a cup of hot coffee cooling down (energy and water molecules spreading to atmosphere), ice melting, and gases mixing. Burning is an exothermic reaction which includes an increase in entropy. It may seem strange that you can measure disorder, but it does enable you to decide whether a reaction is **feasible**.

  The reactants and products involved in a chemical reaction are called a **system**. The rest of the Universe is called the **surroundings**. It is possible to calculate the standard **entropy change $\Delta S^{\ominus}$** of a system during a reaction, as reactants change into products.
- $\Delta H^{\ominus}$ is the standard enthalpy change that accompanies a reaction. **Enthalpy** and **entropy** are **connected** by the equation:
  $\Delta G^{\ominus} = \Delta H^{\ominus} - T\Delta S^{\ominus}$
  $\Delta G^{\ominus}$ is the change in **Gibbs energy**, that is, energy that is **free** to do work. T is temperature in Kelvin.
- Feasible, **spontaneous reactions** (under standard conditions) are indicated when **$\Delta G^{\ominus}$** has a **negative value**. Equilibria lie further to the right with increasingly negative values.

  $\Delta G^{\ominus}$ will have a **negative value** either (i) when **$\Delta H^{\ominus}$ is negative** (the reaction is exothermic which usually means **$T\Delta S^{\ominus}$** is **very small**) or (ii) when **$\Delta H^{\ominus}$** is **positive** (the reaction is endothermic) and the **$T\Delta S^{\ominus}$** term is a **large positive number**, e.g. the reaction between $KHCO_3$ and acid is an endothermic change; it is spontaneous, because it gives off a gas, which involves a large increase in entropy.

  In essence, the requirement for a spontaneous physical change or chemical reaction is that
  $T\Delta S^{\ominus} > \Delta H^{\ominus}$
- Sometimes, reactions **do not occur** when $\Delta G^{\ominus}$ suggests they should. In these cases, you should state that there must be a **high activation energy**, so the rate is low. The reactants are **kinetically stable** due to activation energy (E_{act}) but are **thermodynamically unstable** due to $\Delta G^{\ominus}$.
- Sometimes the $\Delta G^{\ominus}$ term predicts that a reaction is not possible, but it occurs spontaneously **in practice**. In these cases, you should state that **non-standard** conditions are being used. You should work out which conditions are non-standard.
- Often you may rightly suggest a reaction is **feasible** because **$\Delta H^{\ominus}$ is negative**. You may then state that the reaction may **not be feasible** due to a **large negative value of $\Delta S^{\ominus}$**, corresponding to an increase in order.

  Example: Water does not freeze (an **exothermic** change) at room temperature because it involves making ordered ice crystals which is accompanied by a large **entropy decrease**.
- **Calculations** involving entropy and Gibbs energy may look difficult but they are fairly straightforward if you keep cool. Here are some possible types of calculation you might meet:
  (a) The examiner may give you values of $\Delta S^{\ominus}$, T, and $\Delta H^{\ominus}$ and you have to calculate $\Delta G^{\ominus}$ (use $\Delta G^{\ominus} = \Delta H^{\ominus} - T\Delta S^{\ominus}$).
  (b) Conversely, you may be given values for $\Delta G^{\ominus}$, T, and $\Delta H^{\ominus}$ and have to calculate $\Delta S^{\ominus}$.
  (c) You may be given values of $\Delta S^{\ominus}$ for several different reactions and have to calculate $\Delta S^{\ominus}$ for another reaction. You simply construct an entropy cycle diagram involving $\Delta S^{\ominus}$ terms, in the same way that you would construct a Born–Haber cycle using ΔH terms.

TESTS

RECALL TEST

1 Define 'lattice enthalpy'.

(1)

2 Why does CaO have a greater lattice enthalpy than NaCl even though the ionic radii are similar?

(1)

3 Why does MgO have a greater lattice enthalpy than CaO?

(1)

4 Draw a Born–Haber diagram to show how the lattice enthalpy of sodium chloride could be determined. (Use another sheet of paper.)

(4)

5 Why is the theoretical lattice enthalpy of aluminium chloride so different to the calculated (experimental) value for lattice enthalpy?

(1)

6 On a sheet of paper draw an energy cycle to show the connection between lattice enthalpy, hydration enthalpy, and solution enthalpy. What does the sum of the hydration enthalpies equal? (2)

7 Using the terms 'lattice enthalpy' and 'hydration enthalpy', explain why magnesium hydroxide is insoluble and magnesium sulphate is soluble.

(2)

8 Explain the trend in lattice enthalpies of the group 2 oxides from MgO to BaO.

(2)

9 Explain why the group 2 carbonates, with increasing atomic number, become more stable in terms of lattice enthalpy.

(2)

10 What is a strong acid?

(1)

11 Why do all reactions of strong acids with strong alkalis have approximately the same enthalpy of solution?

(1)

12 Why does the reaction of ethanoic acid and sodium hydroxide have a lower enthalpy of neutralisation?

(2)

(Total 20 marks)

CONCEPT TEST

kJ mol^{-1}

$\Delta H_{formation}$(CaO) = –635

$\Delta H_{bond\ enthalpy}$ of O=O = +496

ΔH_a(Ca) = +193

1st electron affinity of O = –142

1st ionisation energy of Ca = +590

2nd electron affinity of O = +844

2nd ionisation energy of Ca = +115

1 a On another sheet of paper, draw a Born–Haber diagram and calculate the lattice enthalpy of calcium oxide. (6)

b Consider a similar diagram for $CaCl_3$. Why is $CaCl_3$ never made?

(2)

c Explain why the theoretical lattice enthalpy of AgI is so different from the experimental lattice enthalpy calculated from a Born–Haber diagram.

____

____ (2)

d Sodium chloride is well known to be soluble.

i Some relevant information is shown right. Calculate the $\Delta H_{solution}$ of sodium chloride.

____

____ (2)

ii Explain why sodium chloride dissolves in terms of the particles, the forces between particles, and the arrangements and kinetic energies of the particles.

____ (4)

	$kJ\,mol^{-1}$
$\Delta H_{hydration}$ of sodium ions	= –406
$\Delta H_{hydration}$ of chloride ions	= –364
lattice enthalpy of sodium chloride	= +771

2 When aqueous KOH reacts with aqueous HCl, both substances are fully ionised. When hydrogen cyanide reacts with KOH the HCN is hardly ionised. The reactions may be expressed as follows:

Reaction I: $KOH(aq) + HCl(aq) \rightarrow KCl(aq) + H_2O(l)$ $\quad \Delta H_n = -57.2\,kJ\,mol^{-1}$

Reaction II: $KOH(aq) + HCN(aq) \rightarrow KCN(aq) + H_2O(l)$ $\quad \Delta H_n = -11.6\,kJ\,mol^{-1}$

Reaction III: $HCN(aq) \rightarrow H^+(aq) + CN^-(aq)$

Assuming aqueous KCI and aqueous KOH are fully ionised, calculate a value for reaction III.

____

____

____ (5)

3 In a blast furnace, iron oxide is reduced by carbon monoxide and by carbon, depending on the temperature.

a Write the equation that links ΔG to temperature.

____ (1)

b The reduction of iron ore could be written thus:

$$FeO(s) + CO(g) \rightarrow Fe(s) + CO_2(g)$$

i Use the data right to calculate the ΔG of reaction at 500 K;

ii Calculate the value for ΔG at 1500 K, given that the reaction $\Delta H = -30.3\,kJ\,mol^{-1}$ and the reaction $\Delta S = -30.0\,J\,mol^{-1}\,K^{-1}$.

____

____ (4)

iii Is the reaction feasible at 500 K or at 1500 K? Explain your answer.

____

____ (2)

	$kJ\,mol^{-1}$
$\Delta G_{formation}(FeO)$	= –231.8
$\Delta G_{formation}(CO)$	= –137
$\Delta G_{formation}(CO_2)$	= –395

iv At what temperature would the reaction change from not being feasible to becoming just feasible? Show any calculations.

____ (2)

(Total 30 marks)

Unit 13 KINETICS I: RATE

- **Rate** is the measure of how fast the **concentration** of a reactant or a product **changes** over time. The units of rate are usually **mol dm^{-3} s^{-1}**.

  The **factors** that **control** rate are temperature, concentration, pressure, catalyst, surface area, light.

- You must remember the **three requirements** for chemical reactions to successfully take place: reactants (atoms, molecules, or ions) must **collide** with each other with the **correct orientation** and with **sufficient energy**.

  Increasing the reactant concentration in solutions increases the reaction rate because there are more reactant particles in a given volume, so they **collide** and react **more frequently**.

  Increasing the pressure of reacting gases increases their concentrations. The reaction rate increases because there are more molecules in a given volume, so they **collide** and react **more frequently**.

  Increasing the surface area of a reactant (e.g. powdering a solid) increases the reaction rate as there is **greater contact** between the reactants, so more collisions take place per second.

- The **energy changes** that happen during a reaction may be expressed as a **reaction profile** (see Figs 13.1 and 13.2).

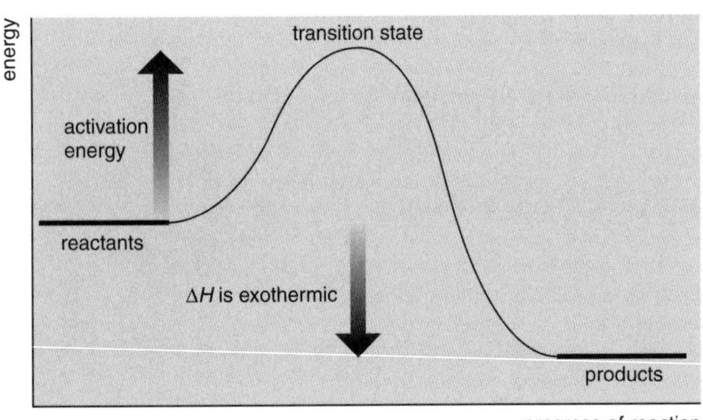

Fig. 13.1 Reaction profile for an exothermic reaction

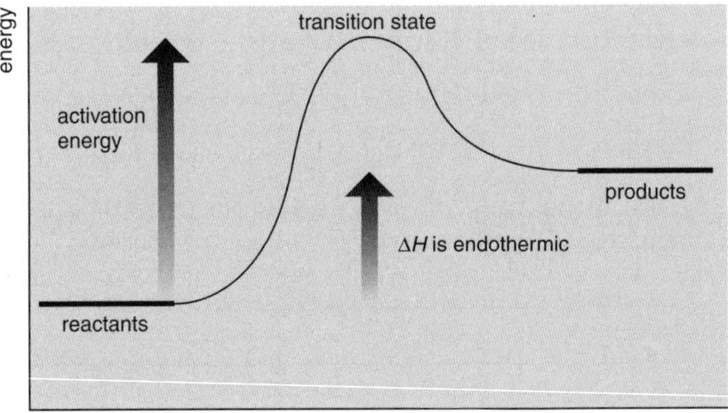

Fig. 13.2 Reaction profile for an endothermic reaction

Primary halogenoalkanes undergo second order nucleophilic substitution, called S_N2. See Fig. 8.5 for mechanism, and Fig. 13.1 for the reaction profile.

- **Activation energy** is the minimum energy required in a collision for the particles to react. Units are J mol^{-1} or kJ mol^{-1} (just like enthalpy). It acts as an **energy barrier** that reactants must overcome if they are to react.

  When molecules react, they form **unstable groups** of atoms called **transition states**. A **trough** at the peak of a reaction profile indicates the existence of an unstable **intermediate compound** (see Fig. 13.3).

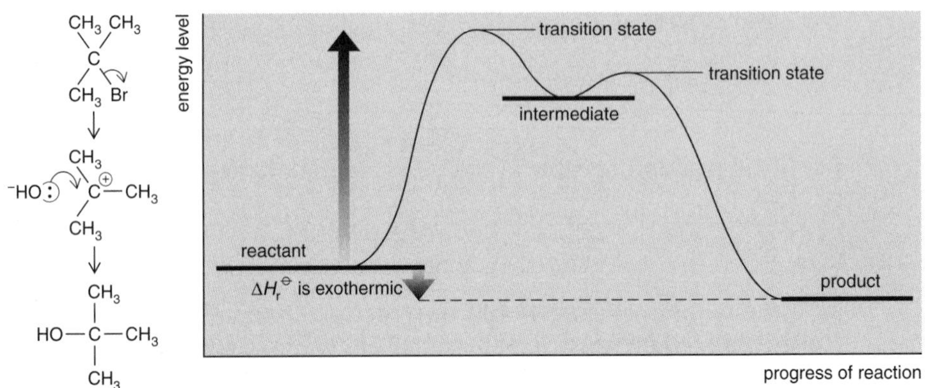

Fig. 13.3 An intermediate forms when a tertiary halogenoalkane undergoes first order nucleophilic substitution (S_N1).

In some reactions, the rate increases when reactants absorb energy from **light** so they can overcome the activation energy barrier.

- The **Maxwell–Boltzmann distribution** of molecular energies gives an instantaneous 'snapshot' (at a specified temperature) of the **proportion** of molecules in a sample that have a given energy (see Fig 13.4).

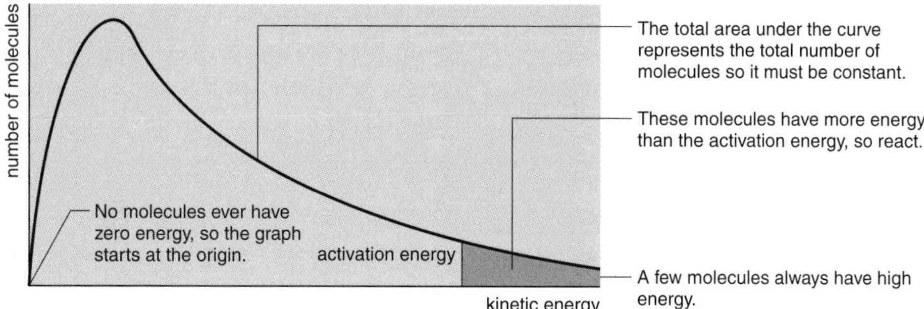
number of molecules
The total area under the curve represents the total number of molecules so it must be constant.
These molecules have more energy than the activation energy, so react.
No molecules ever have zero energy, so the graph starts at the origin.
activation energy
A few molecules always have high energy.
kinetic energy

Fig. 13.4

Note that the molecules with the highest kinetic energy are on the right of the diagram. Look at the **shaded region** to the right of the vertical line representing the **activation energy**. You must understand that all the molecules in this area have **sufficient kinetic energy** to react on collision.

- **Increasing the temperature** increases the reaction rate because molecular **kinetic energy increases** as the temperature rises, which increases the number of molecules that collide with energies greater than the activation energy. Increasing the temperature also increases the **number** of collisions per second (see Fig. 13.5).

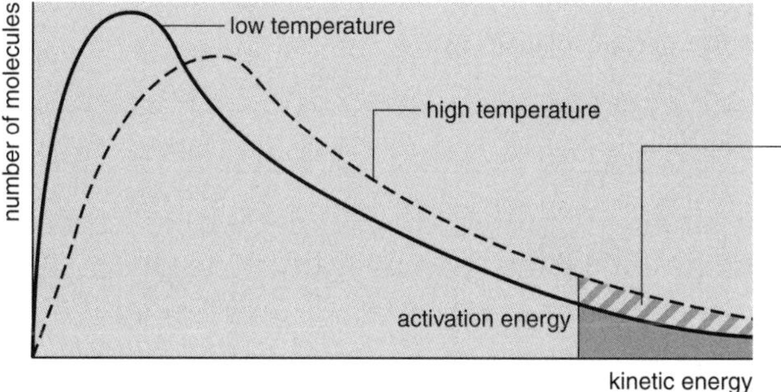
number of molecules
low temperature
high temperature
activation energy
kinetic energy
When the temperature increases, draw the peak of the curve over to the right and down (to keep the area constant).
Here there are more molecules with energy greater than the activation energy, so more are reacting than before.

Fig. 13.5

- **Catalysts** increase the reaction rate without themselves being used up. They function by creating **alternative** reaction routes (via intermediates) with **lower activation energies** (see Fig. 13.6). As a result, there are more molecules present at a given temperature that have sufficient energy to react (see Fig. 13.7).

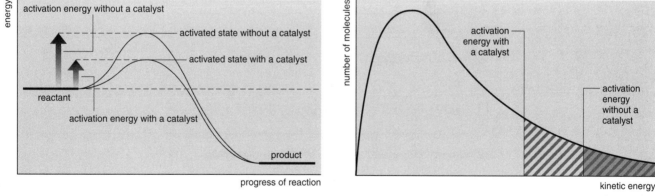

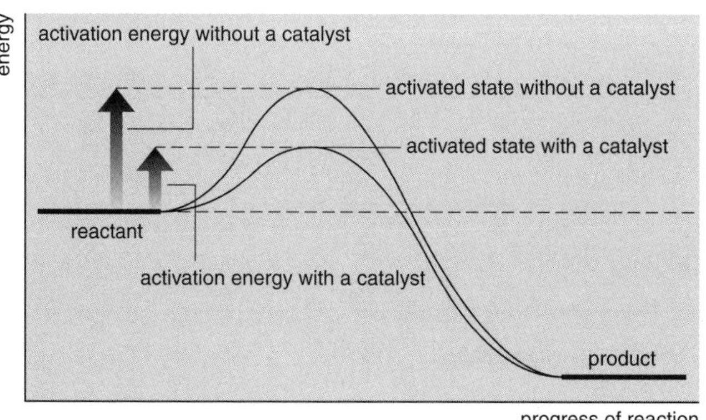
energy
activation energy without a catalyst
activated state without a catalyst
activated state with a catalyst
reactant
activation energy with a catalyst
product
progress of reaction

Fig. 13.6 Reaction profile of a reaction with and without a catalyst

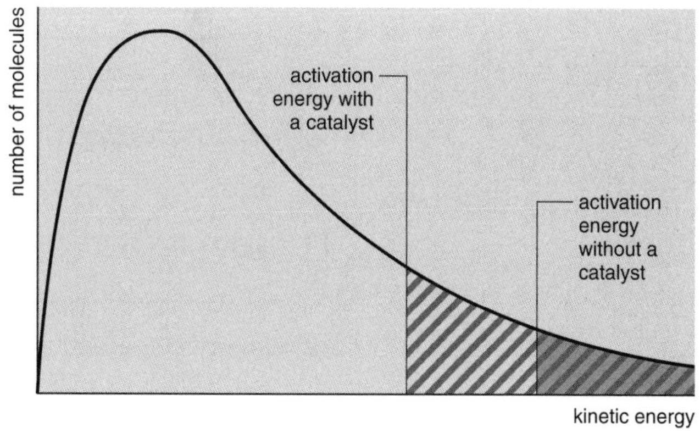

Fig. 13.7

- **Heterogeneous catalysts** are in a **different phase** to the reactants. **Homogeneous catalysts** are in the **same phase** as the reactants. (See unit 16 for examples.)

- Substance **stability** must be discussed in the **context** of a given reaction and with respect to both **thermodynamic** and **kinetic** stability. For example, paper is **thermodynamically unstable** in air with respect to its combustion products; it could oxidise. However, it is **kinetically stable** because the high activation energy means that the oxidation is extremely slow at room temperature.

Unit 13

TESTS

RECALL TEST

1 Define 'rate'. What are the units of rate?

(2)

2 State six factors which influence rate.

(6)

3 What three conditions are required for molecules to react?

(3)

4 How does increasing concentration increase rate?

(2)

5 How does increasing surface area increase rate?

(2)

6 On a piece of paper, draw an energy profile for an endothermic reaction with an intermediate. Label the intermediate, and any transition states. (2)

7 Define 'activation energy'.

(3)

8 On a piece of paper, draw a graph to show why increasing temperature increases rate. (1)

9 Explain why increasing temperature increases rate.

(2)

10 What is a catalyst?

(1)

11 In general, how does a catalyst increase rate?

(2)

12 Give two specific ways in which it does this.

(2)

13 What does it mean when a chemical is 'kinetically stable'?

(1)

14 What does it mean when a chemical is 'thermodynamically stable'?

(1)

(Total 30 marks)

CONCEPT TEST

1 a Define 'activation energy'.

(3)

b On a piece of paper, draw and label the reaction profile for a reaction which is endothermic and forms an intermediate. (3)

c Also, draw and label a graph to explain why rate increases when temperature increases. (3)

d Explain in words why the rate increases when the temperature is increased.

(3)

e What term is used to describe the time taken for the reactant concentration to change?

(1)

2 a On a piece of paper, draw a reaction profile to explain the effect of adding a catalyst. (2)

b Explain in words why the rate increases when a catalyst is added.

(3)

c White phosphorus is unreactive in water, but spontaneously burns in air. Describe these observations in terms of kinetic and thermodynamic stability.

(2)

(Total 20 marks)

Unit 14

KINETICS II: RATE EQUATIONS

- You need to start with some **definitions**:

  Concentration has the units mol dm^{-3}. It is written with square brackets [ ].
  Example: For the decomposition of hydrogen peroxide

  $2H_2O_2(aq) \rightarrow O_2(g) + 2H_2O(l)$

  the concentration of hydrogen peroxide is written as $[H_2O_2(aq)]$.

  The rate equation links rate to the concentrations of the substances that control the reaction rate. **Example:** The rate equation for the decomposition of hydrogen peroxide is

  Rate = $k[H_2O_2(aq)]$ k is the **rate constant** for this expression.

  Order is the sum of the **powers** in the rate equation.
  Example: For the hydrolysis of ethyl bromide (a primary halogenoalkane),

  rate = $k[CH_3CH_2Br(aq)]^1[OH^-(aq)]^1$

  The reaction is 1st order with respect to $[CH_3CH_2Br(aq)]$, 1st order with respect to $[OH^-(aq)]$, and second order **overall** (see Fig 13.3 for mechanism).

- Order must be determined **experimentally**.
  Example: For the iodination of propanone:

  $CH_3COCH_3(aq) + I_2(aq) \rightarrow CH_3COCH_2I(aq) + HI(aq)$

  Experiments show that $[CH_3COCH_3(aq)]$ and $[H^+(aq)]$ influence the rate, but not $[(I_2(aq)]$. Therefore, the rate equation is

  rate = $k[CH_3COCH_3(aq)]^1[H^+(aq)]^1$

  Order is **first order** when the rate is proportional to the substance concentration.
  Example: In reactions where rate = $k[z]^1$

  doubling the concentration of z doubles the rate.

  Order is **second order** if the rate is proportional to the square of the substance concentration.
  Example: in reactions where rate = $k[z]^2$

  doubling (× 2) the concentration of z quadruples (× 4) the rate; tripling (× 3) the concentration of z multiplies the rate by 9.

  Order is **zero order** if the rate is independent of the substance concentration, i.e. varying the substance concentration has no effect on the rate.

  Rate = $k[z]^0$ i.e. rate = k because $[z]^0 = 1$.

- The **slowest step** in a reaction pathway (mechanism) controls the overall reaction rate. This step is called the **rate determining step** or the **rate controlling step**.

- **Half life** is the time taken for concentration of a reactant to fall by a half. See opposite to find out how to determine half life from a graph. Half life has a constant value for first-order reactions.

- During an exam, you may have to explain how to determine order by **experiment**. You must remember to state the three important practical points: (i) that the reaction is **repeated** several times under the **same conditions**; (ii) that the concentration of **one substance only** is changed at a time; and (iii) how the change in reactant concentration is **measured** (see below).

Property change	Instrument
Colour	Colorimeter
pH	pH meter
Electrical conductance	Conductivity meter
Plane of polarisation of light	Polarimeter
Gas volume	Gas syringe
Chemical change	Titration (in exams this one can involve complex explanations)

You must understand the basic principles from the previous unit (13) before studying this one.

Hydrolysis of $(CH_3)_3CBr$ (a tertiary halogenoalkane) is 1st order overall. (See Fig. 13.3 for mechanism.)

You will most likely find that all the reactions you meet in examinations will be zero, first, or second order with respect to each reactant.

Order suggests the **number of particles** involved in the **rate determining step**.

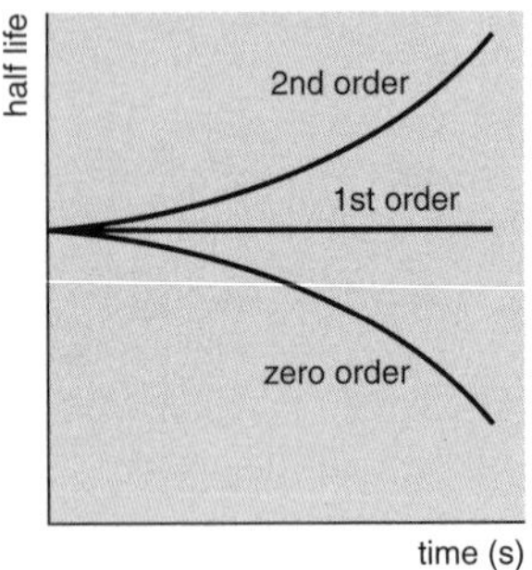

Fig. 14.1

rate constant, $k = Ae^{\frac{-E_a}{RT}}$

E_a = activation energy
R = gas constant
T = temperature (K)
e = exponent
A = Arrhenius constant

$\ln k = \ln A - \frac{E_a}{R}\left(\frac{1}{T}\right)$

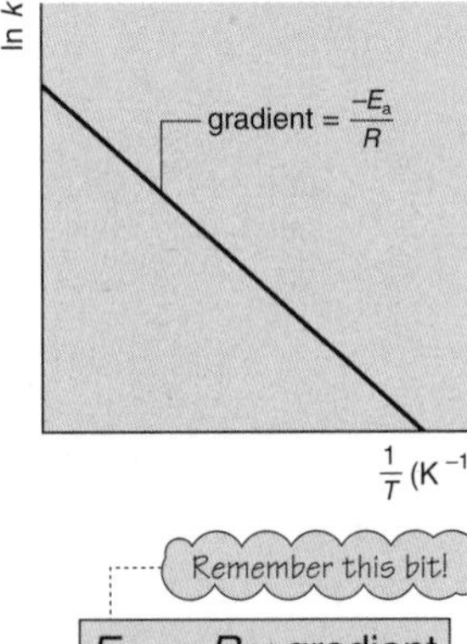

so, $E_a = -R \times$ gradient

Fig. 14.2

The **Arrhenius equation**

$$\ln k = \ln A - \frac{E_a}{RT}$$

links together the **rate constant *k*** and the **activation energy E_a**.

E_a is activation energy, R is the gas constant, T is temperature, and A is the Arrhenius constant.

You do NOT have to remember this equation but you may have to use it to calculate activation energy from given data (see Fig. 14.2).

- You must recognise and understand the **graphs** in Figs 14.1 and 14.3 and know how to use them to work out order, half life, and rate. You may have to work out the gradient on some other sorts of graph – as in diagram (g).

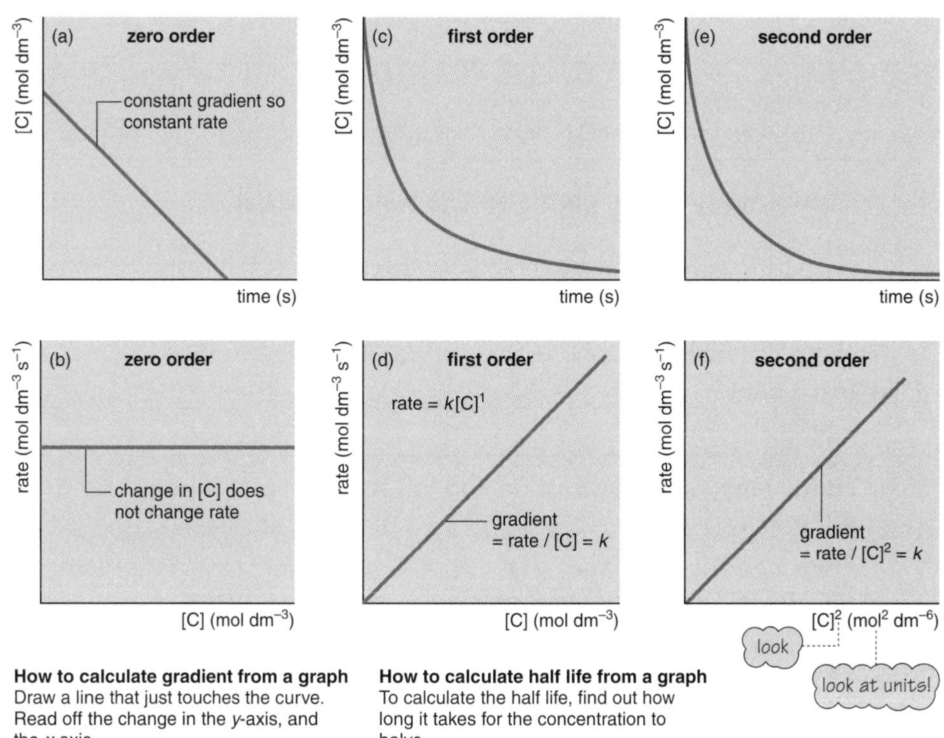

How to calculate gradient from a graph
Draw a line that just touches the curve. Read off the change in the *y*-axis, and the *x*-axis.

$$\text{Gradient} = \frac{\text{change in } y}{\text{change in } x}$$

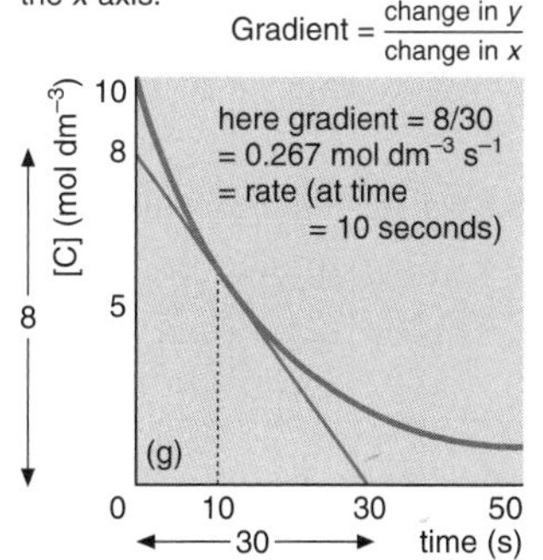

How to calculate half life from a graph
To calculate the half life, find out how long it takes for the concentration to halve.

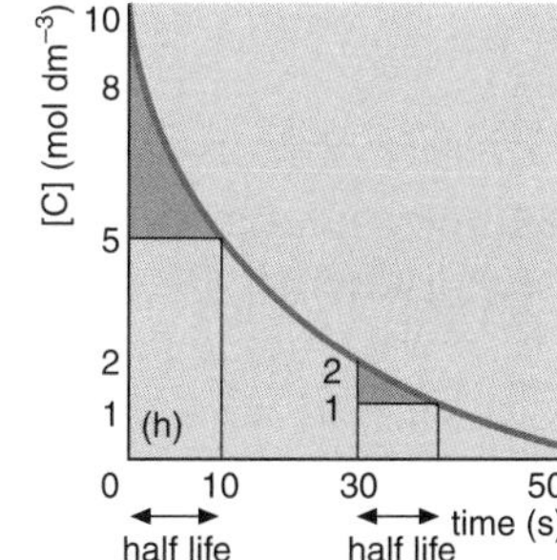

Fig. 14.3

- Fig. 14.4 is a table of concentration data for a reaction.
- Fig. 14.4: The results of these experiments are not easy to interpret. However, you can compare the two experiments in which [q] is constant so you can see the effect of [p]. Experiments G and H show that, when [q] is constant, doubling [p] multiplies the rate by 4, showing that the reaction is **second order with respect to [p]** i.e. rate = $k'[p]^2$. Experiments G and E show that, when [q] is constant, tripling [p] multiplies the rate by 9, confirming that the reaction is **second order with respect to [p]**. To find the order of [q] you should look at experiments H and F, in which [p] is constant. Here, rate doubles when [q] doubles, showing that the reaction is **first order with respect to [q]**, i.e. rate = $k''[q]^1$. The reaction is third order overall i.e. **rate = $k[p]^2[q]^1$**.

Exp.	Rate	[p]	[q]
E	9	3	1
F	8	2	2
G	1	1	1
H	4	2	1

×2 ×2 ×2 ×2 constant

Fig. 14.4

Unit 14

TESTS

RECALL TEST

1 What is the rate equation?

______________________________ (1)

2 What is 'order'?

______________________________ (1)

3 What are the units of k in a second order rate equation? ______ (1)

4 What are the units of k in a first order rate equation? ______ (1)

5 What are the units of k in a zero order rate equation? ______ (1)

6 What does the order suggest?

______________________________ (1)

7 What is 'half life'?

______________________________ (1)

8 State how these physical properties can be measured:

colour ______________________________

pH ______________________________

electrical conductance ______________________________

polarised light ______________________________

gas volume ______________________________ (5)

Rate	[a]	[b]
1	1	1
	2	1
	3	1
	1	2
	1	3

9 Finish the table of rate against [a] and [b], when [a] is first order and [b] is second order. (3)

10 On a separate sheet of paper, draw a graph that summarises rate against half life for 1st, 2nd, and zero order reactants. (3)

11 What does a straight-line graph of concentration against time suggest? What does the gradient tell you?

______________________________ (2)

12 What does a straight-line graph of rate against concentration suggest? What does the gradient tell you?

______________________________ (2)

13 What does a straight-line graph of rate against the square of concentration suggest? What does the gradient tell you?

______________________________ (2)

14 For a graph of ln k against $1/T$, the gradient is –340 K. What must the activation energy be? (The gas constant, $R = 8.314\,\mathrm{J\,K^{-1}\,mol^{-1}}$.)

______________________________ (1)

15 Which graphs give straight lines with reagents that are

a zero order? ______________________________

b first order? ______________________________

c second order? ______________________________ (5)

(Total 30 marks)

CONCEPT TEST

1 The reaction of propanone (CH_3COCH_3) with iodine under acidic conditions was studied. The following data were collected:

$[CH_3COCH_3]$	$[I_2]$	$[H^+]$	Rate
0.1	0.1	0.1	2.0
0.2	0.1	0.1	4.0
0.1	0.2	0.1	2.0
0.1	0.1	0.2	2.0

a Use the data to deduce the order of reaction with respect to these substances:

propanone ______

iodine ______

hydrogen ions ______ (3)

b What part did iodine play in the reaction?

______ (1)

c Write the rate equation for the reaction.

______ (1)

d State the minimum number of steps there must be in the mechanism.

______ (1)

e What must the role of the acid be in the reaction? Explain your answer.

______ (2)

f What is meant by the 'rate determining step'?

______ (2)

g Suggest which chemicals are are involved in the rate determining step.

______ (2)

2 In the reaction between potassium permanganate and ethane-1,2-dioic acid, it is necessary to have acid conditions. This stoichiometric equation describes the reaction:

$$2MnO_4^-(aq) + 5H_2C_2O_4 + 6H^+(aq) \rightarrow 2Mn^{2+}(aq) + 10CO_2(g) + 8H_2O(l)$$

a From the data given, could you write a rate equation for this reaction? Explain your answer.

______ (2)

b How could the progress of the reaction be followed? State simply the method used, and which chemical concentration would be followed.

______ (2)

c State what is meant by the 'half life' of a reaction.

______ (2)

d What does this graph suggest about the order with respect to X?

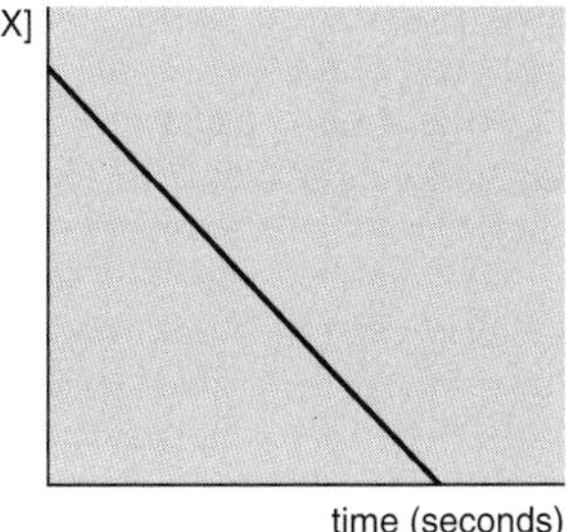

______ (1)

e In a rate equation it is thought that substance Y is first order. Suggest a graph that could be drawn to prove that rate is first order with respect to [Y].

______ (1)

(Total 20 marks)

Unit 15

EQUILIBRIUM I

Reactant substances are those written on the **left** of a chemical equation; **product** substances are written on the **right**.

- A **dynamic equilibrium** develops during all chemical reactions: **concentrations** of substances remain **constant** as reactants change into products (the **forward reaction**) and products revert to reactants (the **backward reaction** – or **reverse reaction**). At **equilibrium**, the **rates** of the forward and backward reactions are equal (see Fig. 15.1).

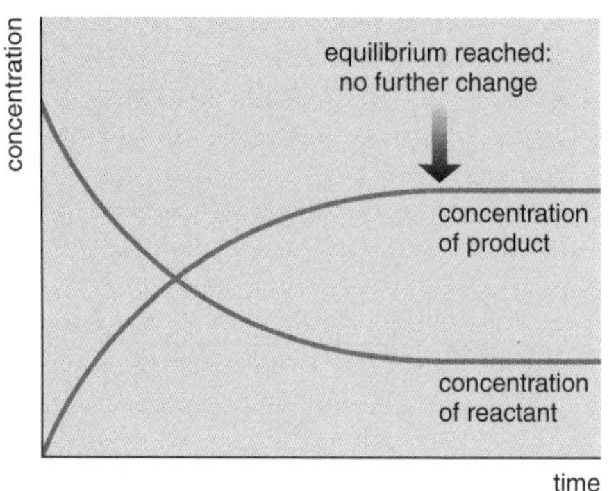

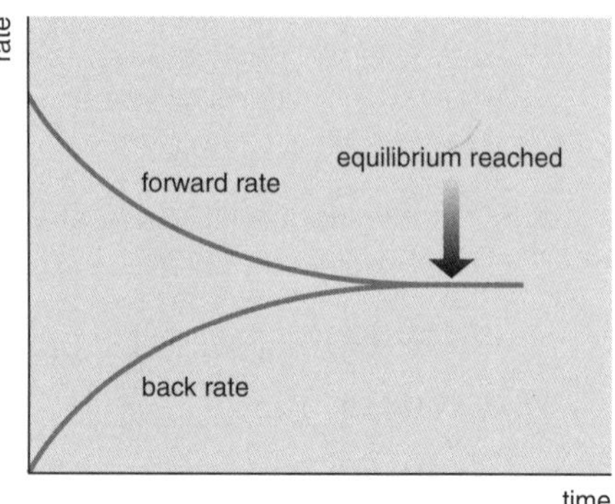

Fig. 15.1

Example: The production of ammonia NH_3 from hydrogen and nitrogen in the Haber process.

$N_2(g) + 3H_2(g) \rightleftharpoons 2NH_3(g)$

At equilibrium, N_2 and H_2 are reacting to make NH_3 at the same rate as NH_3 is decomposing to make N_2 and H_2.

When **product** concentration at equilibrium is **large** compared with reactant concentration, then the equilibrium is said to **lie to the right** and the reaction goes **to completion**. When product concentration at equilibrium is very **small**, then the equilibrium is said to **lie to the left** and the reaction effectively does not take place.

There is a simple relationship at equilibrium between [reactant] and [product] (where square brackets [ ] are used to indicate concentration):

$$\frac{\textbf{[products]}}{\textbf{[reactants]}} = \textbf{a constant } K$$ (called the **equilibrium constant**)

When the **forward** reaction is **exothermic**, the **backward** reaction is **endothermic**. **Example:** The Haber process equilibrium consists of the forward reaction

$N_2(g) + 3H_2(g) \rightarrow 2NH_3(g) \quad \Delta H^\ominus = -92 \text{ kJ mol}^{-1}$

and the backward reaction

$2NH_3(g) \rightarrow N_2(g) + 3H_2(g) \quad \Delta H^\ominus = +92 \text{ kJ mol}^{-1}$.

The equilibrium reaction is written as

$N_2(g) + 3H_2(g) \rightleftharpoons 2NH_3(g)$

and is referred to as an **exothermic reversible reaction**.

- **Le Chatelier's principle** states that if the conditions of a system at equilibrium are changed then the equilibrium position will **shift** to resist the change. You may need to recall this definition, but it is more important to understand how to apply the idea.

If the **concentration** of a substance is **increased**, then the equilibrium will shift to **decrease** its concentration. **Example:** If extra N_2 is added to an equilibrium mixture of H_2, N_2, and NH_3, then the equilibrium shifts to increase the concentration of NH_3 and decrease the concentration of N_2 (and H_2) (see Fig. 15.2).

If a reversible reaction is **exothermic** then an **increase in temperature** will shift the equilibrium to the **left** (in the direction of endothermic change) and the yield will **decrease** i.e. when the temperature is increased by **heat flowing into** the equilibrium mixture, the equilibrium moves in the direction which absorbs heat and lowers the temperature. Note that **increasing** the temperature of the Haber process **decreases** the yield. (see Fig. 15.3.)

The term **reversible reaction** is used to refer to equilibrium mixtures that contain significant amounts of **both** reactants and products.

Example: At equilibrium in the Haber process,

$$K_c = \frac{[NH_3(g)]^2}{[H_2(g)]^3[N_2(g)]}$$

When the value of the equilibrium constant is **large**, the equilibrium position **lies to the right** and products predominate.

When the value of the equilibrium constant is **small**, the equilibrium position **lies to the left** and reactants predominate.

Yield is the percentage proportion of the product in the equilibrium mixture.

The reaction mixture is a **closed system**, which means that there must not be a loss or gain of reactants or products or energy.

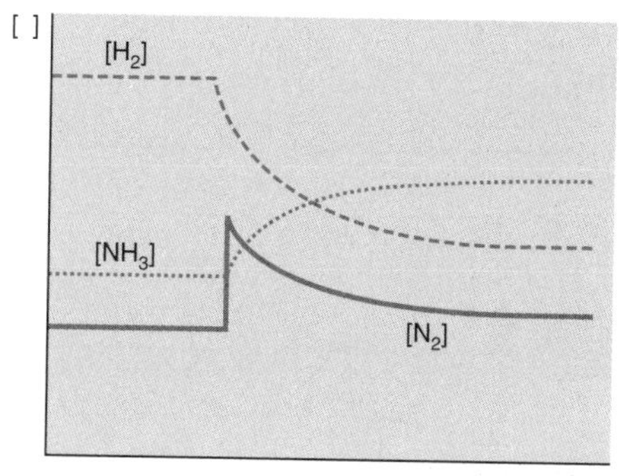

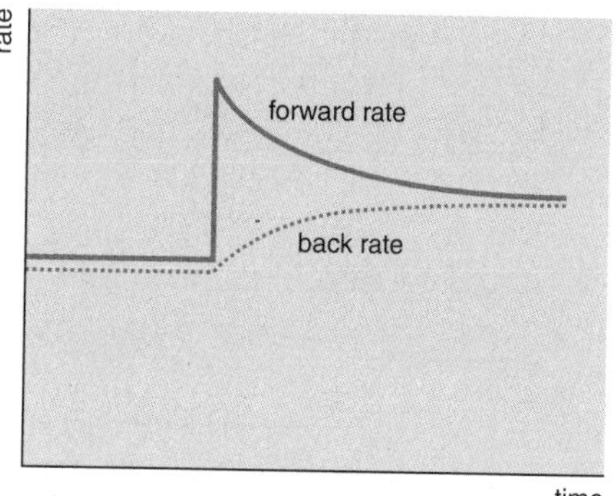

Fig. 15.2

If a reversible reaction is **endothermic** then an **increase in temperature** will shift the equilibrium to the **right** (in the direction of endothermic change) and the yield will **increase** i.e. when temperature is increased, the equilibrium again responds by absorbing heat.

The only condition to change K_c is temperature.

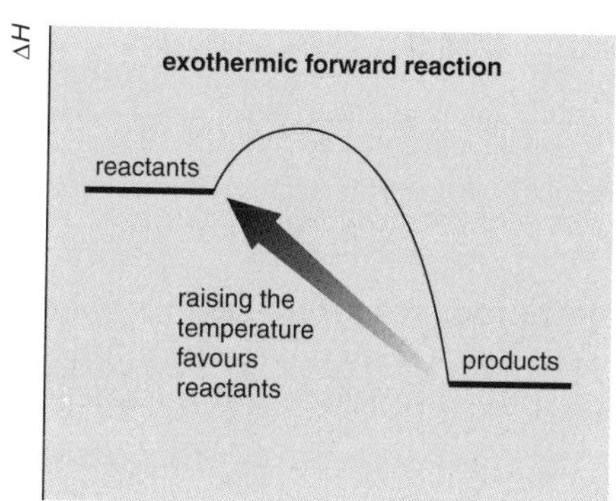

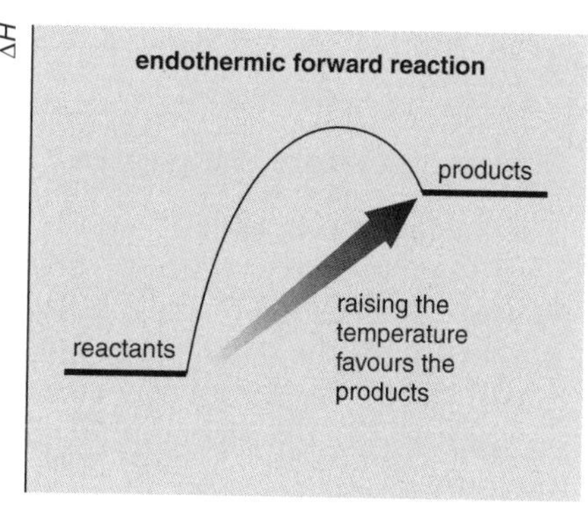

Fig. 15.3

Example: In the roasting of limestone to make lime (used in mortar and for neutralising acid soils), the reaction

$CaCO_3(s) \rightleftharpoons CaO(s) + CO_2(g) \quad \Delta H^{\ominus} = +179 \text{ kJ mol}^{-1}$

is endothermic, so an increase in temperature shifts the equilibrium to the right.

Note if the **temperature increases** the equilibrium position may shift but, at the same time, both forward and backward **rates increase**.

Changing the **pressure** of some gas reactions can alter the **equilibrium position**. **Example:** In the forward reaction of the equilibrium

$N_2(g) + 3H_2(g) \rightleftharpoons 2NH_3(g)$

there is a decrease in the **number of molecules** (4 gas molecules react to make 2 gas molecules). An increase in pressure causes the equilibrium to shift to the right which decreases the total number of molecules in the mixture, which causes the pressure to decrease.

However, in the equilibrium reaction:

$H_2(g) + I_2(g) \rightleftharpoons 2HI(g)$

pressure does not cause the total number of molecules to change because there are the same numbers of gas molecules (two) on each side of the equation. Changing the pressure has no effect on the position of this equilibrium.

- A **catalyst** has no effect on the equilibrium position and so **no effect** on the yield. Catalysts cause the equilibrium position to be reached **more quickly**. They catalyse the rates of the forward and backward reactions to the **same extents**.

Remember:

Catalysts have no effect on Le **Chat**elier.

- As you have seen above, high pressure and low temperature will increase the yield of NH_3 in the **Haber process**. However low temperature causes the reaction rate to be too slow, so a **compromise temperature** must be used (~**500 °C**). Pressures from **200 to 1000 atm.** are used, higher pressures being uneconomic.

TESTS

RECALL TEST

1 What is meant by 'dynamic equilibrium'?

____

____ (1)

2 State Le Chatelier's principle.

____

____ (1)

3 alcohol + acid $\rightleftharpoons$ ester + water (this reaction has ΔH ~ zero)

State which way the position of equilibrium will shift if the following conditions are applied to the reaction (state whether it shifts to the left, right, or does not change):

a [alcohol] is increased ____________________

b [ester] is increased ____________________

c [acid] is decreased ____________________

d temperature is increased ____________________

e a catalyst is added ____________________ (5)

4 $CaCO_3(s) \rightleftharpoons CaO(s) + CO_2(g)$ is an endothermic reaction, used to make basic calcium oxide, used to neutralise acid soils. State which way the equilibrium will shift if the following conditions are applied to the reaction (state whether the reaction equilibrium shifts to the left, right, or does not change):

a pressure is increased ____________________

b temperature is decreased ____________________

c pressure is decreased ____________________

d more calcium carbonate is added ____________________ (4)

5 $H_2(g) + I_2(g) \rightleftharpoons 2HI(g)$ is an exothermic reaction.

State whether the forward rate, backward rate, or yield increase under the following conditions:

a increased temperature ____________________

b increased pressure ____________________

c addition of a catalyst ____________________ (6)

6 Give the stoichiometric (chemical) equation and state the conditions required for these industrial processes:

a production of ammonia ____________________

b production of sulphur dioxide ____________________

c oxidation of ammonia ____________________ (3)

(Total 20 marks)

CONCEPT TEST

1 Consider the equation for the production of ammonia:

$$N_2(g) + 3H_2(g) \rightleftharpoons 2NH_3(g) \quad \Delta H = -92\ kJ\ mol^{-1}$$

One industrial design uses a temperature of 450 °C and a pressure of 300 atmospheres with an iron catalyst.

a Explain why a higher temperature is not used, even though the rate would increase.

___

___ (2)

b Why is a higher pressure not used?

___

___ (2)

c State and explain the effect the catalyst has on the production rate of ammonia.

___

___ (2)

d State and explain the effect adding the catalyst has on the yield at equilibrium.

___

___ (2)

2 This is the important equilibrium equation for the oxidation of ammonia in a nitric acid plant:

$$4NH_3(g) + 5O_2(g) \rightleftharpoons 4NO(g) + 6H_2O(g) \quad \Delta H_r = -905.6\ kJ\ mol^{-1}$$

a What is meant by 'dynamic equilibrium' in the context of this reaction?

___

___ (2)

b State and explain the effect of increasing the temperature on the position of the reaction equilibrium.

___

___ (2)

c Would an increase in pressure increase the yield? Explain your answer.

___

___ (2)

d Why is low pressure used, rather than high pressure?

___

___ (2)

e The NO made is used to make nitric acid. Write an equation for the formation of nitric acid.

___ (2)

f Nitric acid may be used with sulphuric acid. State the conditions required to make sulphur trioxide from sulphur dioxide and oxygen.

___ (2)

(Total 20 marks)

Unit 16

INDUSTRIAL PROCESSES, CATALYSTS, AND THE ENVIRONMENT

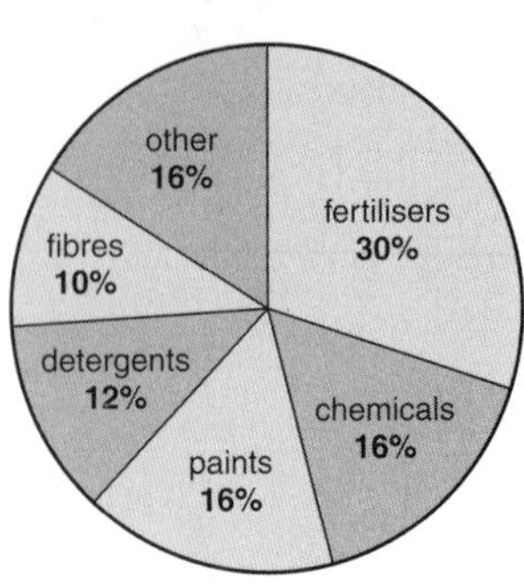

Fig. 16.1 Sulphuric acid uses

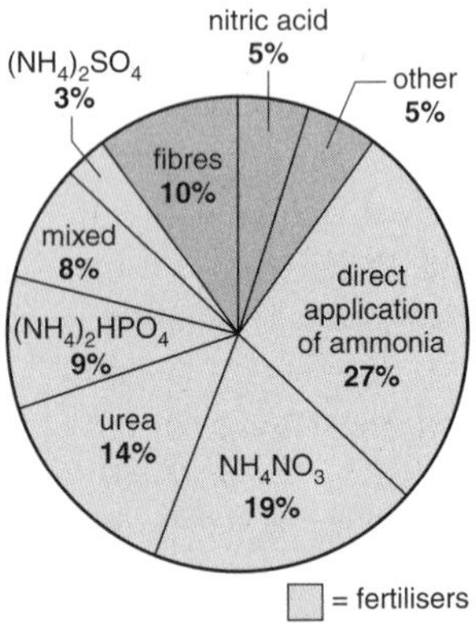

Fig. 16.2 Ammonia uses

Enormous quantities of ammonia, sulphuric acid, and nitric acid are used industrially to make fertilisers, explosives, and polyamides. The base ammonia reacts with sulphuric and nitric acids to make the **fertilisers** ammonium sulphate and ammonium nitrate.

Explosives are made by the **nitration** of organic compounds using concentrated nitric acid. Other nitrated compounds are converted into **polyamides** (e.g. Nylon).

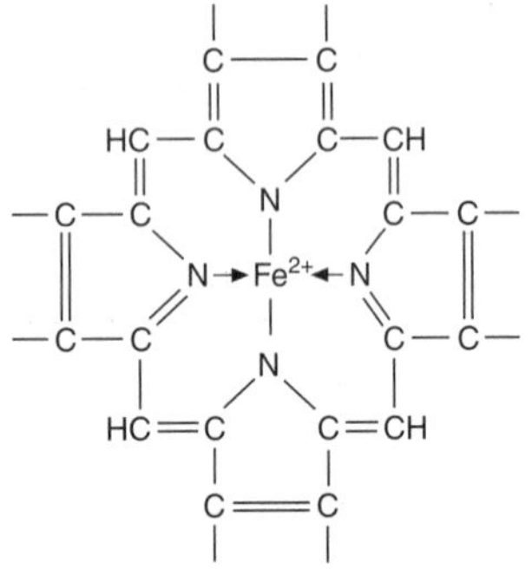

Fig. 16.3 The central part of haem

- To make chemical production on an industrial scale **economically viable**, chemists balance the **reaction kinetics**, **equilibrium**, and **enthalpy** against **economic** and **environmental** factors.
- Sulphuric acid is made from sulphur trioxide which is manufactured by the **Contact process**. Sulphur dioxide and oxygen react to form sulphur trioxide.

  $SO_2(g) + \frac{1}{2}O_2(g) \rightleftharpoons SO_3(g) \quad \Delta H^{\ominus} = -197\ \text{kJ mol}^{-1}$
- The table shows the effects of conditions on rate and SO_3 yield.

	Increase pressure	Increase temperature	Catalyst
Yield	Increases	Decreases	No effect
Rate	Increases	Increases	Increases

The table shows that the use of high pressure, a compromise temperature, and a catalyst will produce the greatest yield. In practice only **1–2 atm. pressure** is required, with a temperature of **450 °C** and a vanadium(V) oxide **V_2O_5 catalyst**. SO_3 and water react together to form sulphuric acid. However, SO_3 forms a stable mist with water rather than dissolving in it. Instead, SO_3 is **dissolved** in **pure H_2SO_4** which is then **diluted**.

$SO_3(g) + H_2SO_4(l) \rightarrow H_2S_2O_7(l)$ (oleum)

$H_2S_2O_7(l) + H_2O(l) \rightarrow 2H_2SO_4(l)$

A huge amount of sulphuric acid is used in industry (see Fig. 16.1).

- **Nitric acid HNO_3** is made from **ammonia** (see Fig. 16.2). The first step is to pass a mixture of NH_3 and air (O_2) over a platinum/rhodium catalyst at 850 °C to make **nitrogen monoxide** (the reaction is exothermic).

  $4NH_3(g) + 5O_2(g) \rightleftharpoons 4NO(g) + 6H_2O(g)$ (yield approx. 96%)

  Pressure, temperature, and catalysts have the same influence on this reaction as in the Contact process. In practice, only sufficient pressure to pump the gases through the catalyst is required. Only initial (electrical) heating of the catalyst gauze is required as the reaction is exothermic. The NO is further oxidised to **nitrogen dioxide** on contact with more air.

  $2NO(g) + O_2(g) \rightarrow 2NO_2(g)$

  Nitric acid forms when the NO_2 (mixed with more air) dissolves in **water**.

  $4NO_2(g) + O_2(g) + 2H_2O(l) \rightarrow 4HNO_3(aq)$
- Petrol is a fraction from the crude oil distillation and made by catalytic cracking, but unimproved petrol would damage car engines by combusting too early, which is called **pre-ignition**. Originally, lead compounds were added to limit pre-ignition, but lead harms the nervous system, and poisons catalytic converters.

  Catalytic converters in car exhaust systems use **Rh** on a ceramic honeycomb (to **economize** on expensive Rh) to catalyse the conversion of pollutants (**CO**, **NO_x**, unburnt **hydrocarbons**) into **CO_2**, **N_2**, and **H_2O**.

  Catalytic **poisons** bind **irreversibly** (permanently) with catalysts and stop them working. **Example: Haem** is a porphyrin complex similar in structure to porphyrin synthetic dyes. This stores oxygen in red blood cells. It is poisoned by CO. (See Fig. 16.3.)

  Branched alkanes, cycloalkanes, and arenes are used in petrol to encourage efficient combustion. These chemicals are made by catalytic cracking, catalytic reforming, and isomerisation. Also methanol is added to improve combustion.

As crude oil becomes scarce and expensive, it will be replaced by new fuels. Countries without sources of crude oil use alcohol, produced from plants, in their 'biofuels'. Ethanol or methanol could replace petrol, though ethanol combustion produces less energy per kilogram. Cheaper to run than petrol, gaseous fuels are already used, but do require bulky high-pressure fuel tanks. Hydrogen may be the fuel of the future because it is non-polluting, but it requires heavy fuel tanks. Solar, wind, tidal, and nuclear energy could supply the energy required to manufacture hydrogen.

- **Homogeneous catalysts** are in the **same** phase as the reactants. **Example:** Fe^{2+}(aq) catalyses the redox reaction between peroxodisulphate and iodide ions:

  $2I^-(aq) + S_2O_8^{2-}(aq) \rightarrow I_2(aq) + 2SO_4^{2-}(aq)$

- **Heterogeneous catalysts** are in a **different** phase to the reactants. **Example:** Nickel catalysing the reaction between ethene and hydrogen to make ethane. The nickel acts as a surface catalyst, **adsorbing** the reactant gas molecules so they are **correctly aligned** for easy reaction, and then **desorbing** the product. Tungsten adsorbs too strongly and silver too weakly, so they are not suitable. Nickel and platinum are ideal.

  In industry, $AlCl_3$ with HCl or a phosphoric acid catalyst helps **ethane** to react with **benzene** to make **ethylbenzene**, which is then **dehydrogenated** to make **phenylethene** (used to manufacture polystyrene).

  $$C_6H_6 + CH_2{=}CH_2 \xrightarrow{AlCl_3/HCl} C_6H_5CH_2CH_3$$

  $$C_6H_5CH_2CH_3 \xrightarrow{Cr/Al_2O_3} C_6H_5CH{=}CH_2 + H_2$$

LRF (lead replacement fuel) has replaced leaded petrol.

LPG (liquid petroleum gas) contains butane.

Natural gas contains methane.

Gasohol is an ethanol–petrol mix.

- Biological catalysts called **enzymes** are increasingly used to make useful chemicals. **Example:** Ethanol is made by **fermentation** of warm sugar solution, using enzyme-containing yeast. Industrially, ethanol is made by **hydrating ethene** over a catalyst of phosphoric acid at 300 °C and 70 atm.

Fermentation uses renewable resources but the product is more expensive and impure.

- The relatively strong C-Cl bond in **organo-chlorine** compounds such as insecticides makes them stable and **persistent**.

Many halogenoalkanes are **toxic** by poisoning enzymes.

- **Chlorofluorocarbons** (CFCs) are compounds in which some or all of the hydrogen atoms are replaced by chlorine and fluorine atoms. They are useful as refrigerants, aerosol propellants, plastics foaming agents, in dry cleaning, and for degreasing metal. The **stability** of CFCs allows them to survive into the **ozone layer**, where UV radiation breaks them down. Free radicals are released which **deplete** the ozone, allowing more ultraviolet light to reach the surface and increasing skin cancers and lowering crop yields. Modern non-toxic and non-flammable **replacements** are the more expensive **hydrofluorocarbons** (HFCs) such as CH_2FCF_3, which produce almost zero ozone depletion.

- Transition metals have **catalytic properties** due to their **variable oxidation states**. **Example:** In the Contact Process, V_2O_5 catalyses the overall reaction

  $2SO_2(g) + O_2(g) \rightleftharpoons 2SO_3(g)$

  The reaction happens in two stages:

  (i) $SO_2 + V_2O_5 \rightarrow SO_3 + 2VO_2$

  (ii) $2VO_2 + \frac{1}{2}O_2 \rightarrow V_2O_5$

You must also remember the following catalysts: Haber process (ammonia) = **Fe**; Ostwald process (nitric acid) = **Rh/Pt**; hydrogenation of >C=C< in vegetable oil to C-C (margarine) = **Ni**.

Unit 16

TESTS

RECALL TEST

1 Which five factors decide whether reaction conditions are economically viable?

(5)

2 State five chemical factors that influence the rate of reaction.

(5)

3 Which four factors influence the yield?

(4)

4 If a reaction is exothermic, does high or low temperature increase the yield?

(1)

5 How do economic factors influence the choice of temperature and pressure?

(1)

6 Give the equation for the production of sulphur trioxide.

(1)

7 For this reaction, state the effect of decreasing pressure on the yield and the rate.

(1)

8 Also, state the effect of decreasing temperature on the yield and the rate.

(1)

9 Also, state the effect of adding a catalyst on the yield and the rate.

(1)

10 State the actual conditions used to produce sulphur trioxide.

(2)

11 Sulphur trioxide is converted into sulphuric acid by what?

(1)

12 Give the equation for the oxidation of ammonia to produce nitrogen monoxide.

(1)

13 Give the equation for the oxidation of nitrogen monoxide.

(1)

14 Ammonia, sulphuric acid, and nitric acid are required in large quantities to make what?

(3)

15 Give an example of a nitrogen-containing fertiliser.

(1)

16 Why were lead compounds added to petrol in the past?

(1)

17 Give two different reasons for the phasing out of leaded fuels.

(2)

18 What is the function of catalytic converters?

(3)

19 What metal is used in catalytic converters?

____ (1)

20 How do poisons stop catalysts working?

____ (1)

21 How is ethanol made industrially?

____ (1)

22 Give a use for ethanol.

____ (1)

23 Give a different use for methanol.

____ (1)

(Total 40 marks)

CONCEPT TEST

1 With crude oil prices increasing, methanol production could be developed to replace petrol, or methanol could be used to lessen pre-ignition in petrol.

One way of producing methanol is from the exothermic reaction of carbon monoxide and hydrogen, which are produced from coal and water. This methanol reaction is of industrial importance:

$CO(g) + 2H_2(g) \rightarrow CH_3OH(g)$ using a $ZnO/CrO_3(s)$ catalyst.

a What is pre-ignition, and why is it costly?

____

____ (2)

b Why is a catalyst used to produce methanol?

____

____ (2)

c What would be the effect of increasing the pressure on the reaction?

____

____ (2)

d How would increasing the temperature influence the reaction?

____

____ (2)

e Why would widespread use of methanol as a fuel improve the environment?

____

____ (2)

f How do catalytic converters improve the environment?

____

____ (3)

g Is the catalyst homogeneous or heterogeneous? Give your reasons.

____

____ (2)

(Total 15 marks)

Unit 17 EQUILIBRIUM II: CALCULATIONS

You may need to review the contents of unit 15.

Remember that square brackets [ ] stand for the concentration of a substance in mol dm^{-3}.

The **equilibrium law** relates the concentrations of the reactants and products in an equilibrium mixture.

When the equilibrium constant does not have units, you should write (**no units**).

- K_c is the equilibrium constant for a reversible reaction written in terms of the **concentrations** of the reactants and products. The calculations involve the concentration of each substance raised to the same power as the number of moles shown in the balanced equation.
  Example: The production of ammonia NH_3 from hydrogen and nitrogen in the Haber process.
  $N_2(g) + 3H_2(g) \rightleftharpoons 2NH_3(g)$
  At equilibrium, $K_c = \dfrac{[NH_3(g)]^2_{eq}}{[H_2(g)]^3_{eq}[N_2(g)]_{eq}}$
  In this example, K_c has the units $\dfrac{(\text{mol dm}^{-3})^2}{(\text{mol dm}^{-3})^3(\text{mol dm}^{-3})}$
  i.e. $\text{mol}^{-2}\ \text{dm}^6$
- You must remember that K_c is quoted at a **constant temperature**. Changing the temperature changes its value and the position of the equilibrium (so changes the proportions of reactants and products in the equilibrium mixture).
- **Ignore solids** when writing K_c equations. The density of a solid has a **fixed value**: therefore, the concentration of a solid is a **constant**.
  Example: Roasting limestone to make quicklime involves two solids and a gas.
  $CaCO_3(s) \rightleftharpoons CaO(s) + CO_2(g)$
  $K_c' = \dfrac{[CaO(s)]_{eq}[CO_2(g)]_{eq}}{[CaCO_3(s)]_{eq}}$ so $K_c = [CO_2(g)]_{eq}$
  because the values of the concentrations of the solids are included in the value of the equilibrium constant.

Example: What is the mole fraction of alcohol in 4% aqueous alcohol?
(i) Of 100 g of solution 4 g are alcohol.
(ii) Moles of water = $\dfrac{96}{18} = 5.333$ moles
(iii) Moles of alcohol = $\dfrac{4}{42} = 0.095$ moles
(iv) Total moles = 5.333 + 0.095 = 5.428 moles
(v) Mole fraction = $\dfrac{0.095}{5.428} = 0.018$

- Do **not include** water in the equilibrium expression for reactions that happen in **aqueous** solution; even when a reactant or a product, its concentration effectively remains constant. **Include** the concentration of water when it is a reactant or a product in **non-aqueous** reactions.
- You may have to calculate K_c from given values of **equilibrium concentrations**.
  At equilibrium $[CH_3COOH] = 0.01\ \text{mol dm}^{-3}$, $[CH_3CH_2OH] = 0.15\ \text{mol dm}^{-3}$
  $[CH_3COOCH_2CH_3] = 0.018\ \text{mol dm}^{-3}$, $[H_2O] = 0.02\ \text{mol dm}^{-3}$
  $K_c = \dfrac{0.018 \times 0.02}{0.01 \times 0.15} = 0.24$ no units
- You may have to calculate K_c from given values of **initial** and **equilibrium** concentrations (see Fig. 17.1).

$CO(g) + H_2(g) = C(s) + H_2O(g)$ at 600 °C (volume = 2.0 dm^3)
$K_c = \dfrac{[H_2O]}{[CO][H_2]}$
Initially [CO] = 2.0 moles, $[H_2]$ = 2.0 moles.
At equilibrium $[H_2O]$ = 0.40 moles.

	CO	H_2	H_2O	C
initial moles	2	2	0	0
eqm moles	2 – 0.4 = 1.6	2 – 0.4 = 1.6	0.4	0.4
eqm conc. mol dm^{-3}	$\dfrac{1.6}{2.0} = 0.8$	$\dfrac{1.6}{2.0} = 0.8$	$\dfrac{0.4}{2.0} = 0.2$	$\dfrac{0.4}{2.0} = 0.2$

$\therefore\ K_c = \dfrac{0.2}{0.8 \times 0.8} = 0.31\ \text{mol}^{-1}\text{dm}^3$

Fig. 17.1

- You may have to work out equilibrium concentrations from given values of **initial concentrations** of reactants and K_c (see Fig. 17.2).

$A(aq) \rightleftharpoons B(aq)$

Initially there is 15 mol of A, and $K_c = 2.0$.

$$K_c = \frac{[B]}{[A]} = 2.0$$

	[A]	[B]
initial	15	0
eqm	15 – *x*	*x*

$$2 = \frac{x}{15 - x} \quad \therefore 30 - 2x = x$$
$$\therefore 3x = 30$$
$$\therefore x = 10$$

Therefore at eqm. [A] = 15 – 10 = 5 mol dm^{-3}

[B] = 10 mol dm^{-3}

Take care to **set out** your calculations carefully. In fact the examiners expect you to lay out the calculation in this way. Also, when you are stuck, the method helps you think logically.

Fig. 17.2

- **Mole fraction** = $\dfrac{\textbf{moles of substance}}{\textbf{total moles present}}$
- **Partial pressure of a gas = mole fraction × total pressure.**
  The partial pressure of a gas is indicated by placing the letter 'p' before its chemical symbol. e.g. pN_2O_4
  For a mixture of gases: **total pressure = sum of all partial pressures**
- K_p is the equilibrium constant for a reversible reaction that takes place in the gas phase, written in terms of the **partial pressures** of the reactants and products. It is calculated in a similar way to K_c.
  Example: The Haber process.
  $$N_2(g) + 3H_2(g) \rightleftharpoons 2NH_3(g) \quad K_p = \frac{pNH_3^2}{pN_2 \times pH_2^3}$$
  The units in this case are kPa^{-2} (or atm^{-2}, depending on the partial pressure units).
- You may have to calculate K_p given **equilibrium mole** values for the reactants.
- **Example:** The hydrogen iodide equilibrium.
  $2HI(g) \rightleftharpoons H_2(g) + I_2(g)$
  Starting with 2.0 moles of HI, 1.6 moles are present at equilibrium at 430 K and 2.0 atm pressure.
  $$K_p = \frac{pH_2 \times pI_2}{pHI_2} = \frac{0.2 \times 0.2}{1.6} = 0.064 \text{ (no units)}$$
  Short cut: when there are no units, total pressure terms **cancel** and total volume terms **cancel**. You can calculate K_p from the concentration terms because under these circumstances, K_c and K_p have the same numerical values.
- **Example:** The synthesis of ammonia.
  $N_2(g) + 3H_2(g) \rightleftharpoons 2NH_3(g)$
  Starting with 1.0 moles of N_2 and 3.0 moles of H, there were 0.2 moles of ammonia present in the equilibrium mixture at 100 atm and 800 K. Calculate K_p.
  The remainder of the calculation is set out in Fig. 17.3.

Example: What is the partial pressure of N_2O_4 when the mole fraction of N_2O_4 is 0.24 and the total pressure is 200 kPa?

Partial pressure = 0.24 × 200 = 48 kPa

Just like K_c, the value of K_p is **temperature dependent**.

	N_2	H_2	NH_3
initial moles	1	3	0
equilibrium moles	1 – 0.1 = 0.9	3 – 0.3 = 2.7	0.2
mole fraction	$\frac{0.9}{3.8} = 0.24$	$\frac{2.7}{3.8} = 0.71$	$\frac{0.2}{3.8} = 0.05$
partial pressure	0.24 × 100 = 24.0	0.71 × 100 = 71.0	0.05 × 100 = 5.00

Total moles = 3.8
Total pressure = 100 atm

Fig. 17.3

$$K_p = \frac{(5.00)^2}{24.0 \times (71.0)^3} = 2.91 \times 10^{-6} \quad \text{units} = \frac{\text{atm}^2}{\text{atm} \times \text{atm}^3} = \frac{1}{\text{atm}^2} = \text{atm}^{-2}$$

TESTS

RECALL TEST

1 State Le Chatelier's principle.

______________________________ (2)

2 The units of concentration are ______________________________ (1)

3 a Give the expression for K_c for this reaction: $2SO_2(g) + O_2(g) \rightleftharpoons 2SO_3(g)$.

______________________________ (1)

b What would the units be for this K_c?

______________________________ (1)

4 A value of K_c is quoted for a ______________ temperature. (1)

5 The decomposition of calcium carbonate is $CaCO_3(s) \rightleftharpoons CaO(s) + CO_2(g)$.

a For this reaction write the equation for K_c.

______________________________ (1)

b Write the units for this K_c. ______________________________ (1)

6 What is the mole fraction of glucose in a solution containing 5 g glucose in 100 g water?

____

______________________________ (3)

7 What does partial pressure equal?

______________________________ (1)

8 What is the partial pressure of oxygen if 28 g of oxygen is mixed with 12 g of helium when the total pressure is 50 kPa?

____

______________________________ (3)

9 If pAr = 1 kPa, pO_2 = 10 kPa, and pN_2 = 80 kPa, what is the total pressure?

______________________________ (1)

10 a Write the equation for K_p for the reaction $2SO_2(g) + O_2(g) \rightleftharpoons SO_3(g)$.

______________________________ (1)

b State the units for K_p (when Pa are the units of pressure).

______________________________ (1)

11 Why must the values for K_c and K_p be stated for a specified fixed temperature?

______________________________ (1)

12 The only thing to change the value of K_c or K_p is ______________ (1)

(Total 20 marks)

CONCEPT TEST

1 The ester ethyl ethanoate may be produced by this reaction:

$$CH_3CH_2OH + CH_3COOH \rightleftharpoons CH_3COOCH_2CH_3 + H_2O$$

a Given that initially one mole each of CH_3CH_2OH and CH_3COOH were mixed together, and that at equilibrium 0.8 moles of CH_3COOH was found to be present, calculate the value of K_c.

(6)

b Often when making the ester in the laboratory, concentrated sulphuric acid is added. How does this increase the yield of ester?

(3)

c A catalyst may be added to the mixture. Explain the effect this would have on the yield at equilibrium.

(2)

2 This question concerns the efficiency and economics of the production of synthesis gas represented by this reaction:

$$CH_4(g) + H_2O(g) \rightleftharpoons CO(g) + 3H_2(g) \qquad \Delta H = +523 \text{ kJ mol}^{-1}$$

a Describe and explain the effect of increasing the temperature on the yield.

(2)

b Give an expression for K_p for this reaction.

(1)

c If the units of pressure are Pa, what are the units of K_p?

(1)

d State and explain the effect of increasing the pressure on the equilibrium constant K_p.

(2)

3 Hydrogen iodide will decompose when heated. This may be represented by $2HI(g) \rightleftharpoons H_2(g) + I_2(g)$.

a Calculate the partial pressure of HI after decomposition.

(3)

b If the hydrogen iodide is heated to 760 K at 100 kPa then 15% is found to be decomposed. Calculate the value of K_p.

(5)

c What effect would doubling the pressure have on the amount decomposed? Explain your answer.

(3)

d **i** Give an expression for K_p for $BaCO_3(s) \rightleftharpoons BaO(s) + CO_2(g)$.

(1)

ii If the amount of solid $BaCO_3$ were increased, what effect would this have on the partial pressure of $CO_2(g)$ in an enclosed container?

(1)

(Total 30 marks)

Unit 18

EQUILIBRIUM III: ACID–BASE

$HCl(aq)$, $H_2SO_4(aq)$, and $HNO_3(aq)$ are strong acids; NaOH is a strong base.

$H_2CO_3(aq)$, $H_2SO_3(aq)$, and $HNO_2(aq)$ are all weak acids; $NH_3(aq)$ and $CH_3NH_2(aq)$ are weak bases.

The conjugate base of ethanoic acid is the ethanoate ion $CH_3CO_2^-(aq)$.

The conjugate acid of ammonia is the ammonium ion $NH_4^+(aq)$.

To **convert** $[H^+]$ to pH using a calculator: (i) type in the value of $[H^+(aq)]$; (ii) press the log button; (iii) multiply by –1 (in your head or using the keypad).

- **The Brønsted–Lowry theory:** Acids are proton **donors** and bases are proton **acceptors. Example:**

  $HCl + H_2O \rightarrow H_3O^+ + Cl^-$

  Acid Base Acid Base

  The HCl and Cl^- are a **conjugate pair**. Also H_2O and H_3O^+ are a conjugate pair.

  A **conjugate base** is made when its **conjugate acid** loses an H^+ ion.

  Strong acids and bases are fully ionised in water.

  Weak acids and bases are partly ionised in water. Look for the $\rightleftharpoons$ sign.
  Example:

  $NH_3 + H_2O \rightleftharpoons NH_4^+ + OH^-$

  Base Acid Acid Base

  Here the NH_3 and NH_4^+ are paired. H_2O and OH^- are a conjugate pair.

  Every acid has a corresponding **conjugate base** which is formed when the acid loses an proton. A conjugate base can accept a proton to regenerate the original undissociated acid.

  Every base has a corresponding **conjugate acid** which is formed when a base accepts a proton. A conjugate acid can lose a proton to regenerate the original undissociated base.

  pH is a measure of the concentration of $H^+(aq)$. **$pH = -\log_{10} [H^+(aq)]$.**

  Example: If a solution of ethanoic acid has $[H^+(aq)] = 1 \times 10^{-5}$ mol dm^{-3} then pH = 5.0.

  A **weak acid** such as ethanoic acid is partially dissociated:

  $CH_3COOH(aq) \rightleftharpoons CH_3CO_2^-(aq) + H^+(aq)$

  (This equation is equivalent to the more complex one above.)

- For any weak acid HA: $HA(aq) \rightleftharpoons H^+(aq) + A^-(aq)$

  This reversible reaction gives the equilibrium expression

  $$K_a = \frac{[H^+(aq)][A^-(aq)]}{[HA(aq)]}$$

  K_a is the **acid dissociation constant**. The greater it is, the stronger the acid.

  Example: A solution is prepared by dissolving 0.01 moles of pure acid in water to make 1.0 dm^3 of solution; $[H^+(aq)] = 0.001$ mol dm^{-3}. Use the tabular method shown in Fig. 18.1 to calculate the K_a value.

Fig. 18.1 (concentrations given in mol dm^{-3})

	[HA]	[H$^+$]	[A$^-$]
initial	0.01	0	0
equilibrium	0.01 – 0.001 = 0.009	0.001	0.001

$$K_a = \frac{0.001 \times 0.001}{0.009} = 1.11 \times 10^{-4}\ \text{mol dm}^{-3}$$

  Example: A solution is prepared by dissolving 0.1 moles of benzoic acid in water to make 1.0 dm^3 of solution. Given that K_a for benzoic acid is 6.3×10^{-5} mol dm^{-3}, use the method given in Fig. 18.2 to calculate the pH.

Fig. 18.2 (concentrations given in mol dm^{-3})

	[benzoic acid]	[H$^+$]	[benzoate ions]
initial	0.1	0	0
equilibrium	0.1 – x = 0.1	x	x

so $6.3 \times 10^{-5} = \frac{x^2}{0.1}$

$\therefore$ pH = 2.6

  As x is very small 0.1 – x approximately equals 0.1.

- K_a may be converted to pK_a to give numbers that are more convenient to manipulate and enable acid strengths to be compared more easily.

  $pK_a = -\log K_a$

- You must remember that, when an acid is **50% dissociated**, $[HA(aq)] = [A^-(aq)]$. Therefore, $K_a = [H^+(aq)]$, so **pK_a = pH**. (See indicators, below).

- **Water dissociates** and sets up the following equilibrium:

  $H_2O(l) \rightleftharpoons H^+(aq) + OH^-(aq)$

  The dissociation is very slight so the concentration of water is unaffected.

  $\boldsymbol{K_w = [H^+(aq)][OH^-(aq)]}$

  K_w is the **ionic product of water** (the equilibrium constant for the ionization of water) and has the value $1.0 \times 10^{-14}\ mol^2 dm^{-6}$ at room temperature.

  $pK_w = -\log K_w$, so $pK_w = 14.0$

$pOH = -\log [OH^-(aq)]$

Example: Calculate the pH of a 0.01 mol dm^{-3} solution of NaOH(aq).

NaOH is a strong base so it is fully ionized.

So $[OH^-(aq)] = [NaOH(aq)] = 0.01$.

$pOH = -\log 0.01 = 2.0$

$pH = 14.0 - pOH$
$= 14.0 - 2.0$
$= 12.0$

- **Buffer solutions** resist attempts to change their pH by the addition of (small amounts of) acid or base.

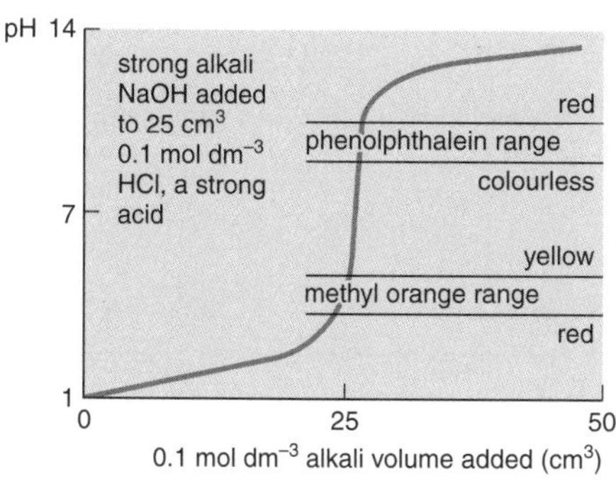

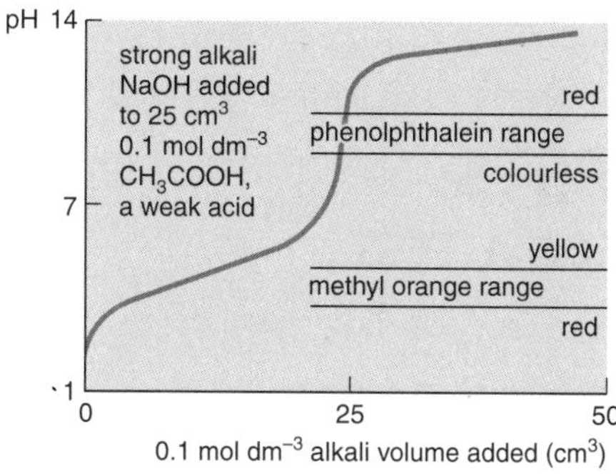

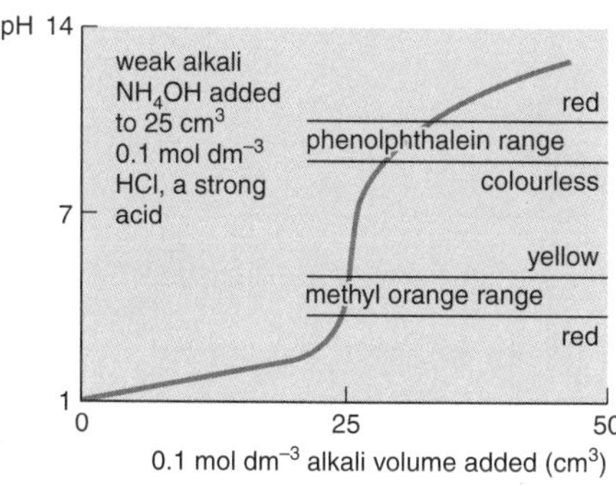

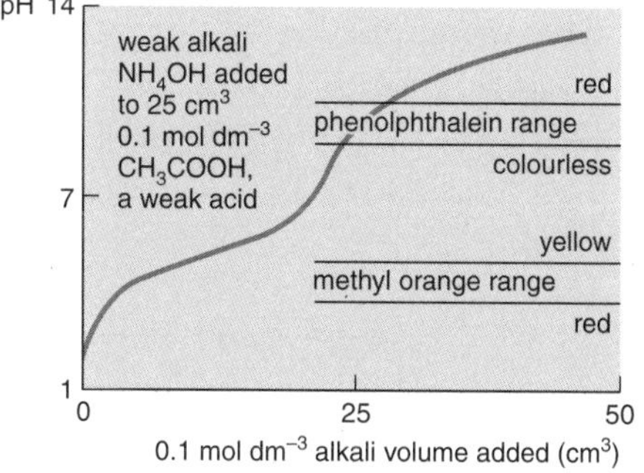

Fig. 18.3 Using a **pH** meter to monitor changes in pH during a normal acid–base titration will produce the **titration curves** shown here.

- **Acid buffers** are made from a mixture of a **weak acid** and a **salt of a weak acid**. **Example:** Ethanoic acid with sodium ethanoate. The mixture produces a high concentration of both $\boldsymbol{CH_3COOH(aq)}$ and $\boldsymbol{CH_3CO_2^-(aq)}$. Added $\boldsymbol{H^+(aq)}$ is removed from solution as it combines with $\boldsymbol{CH_3CO_2^-(aq)}$.

  $CH_3COO^- + H^+ \rightarrow CH_3COOH$

  Added $\boldsymbol{OH^-(aq)}$ is removed from solution as it combines with $H^+(aq)$ from $CH_3COOH(aq)$:

  $CH_3COOH + OH^- \rightarrow CH_3COO^- + H_2O$

  Basic buffers are made from a mixture of a **weak base** with a **salt of the weak base**.

  Example: Aqueous ammonia with ammonium chloride.

  $NH_3(aq) + H_2O(l) \rightleftharpoons NH_4^+(aq) + OH^-(aq)$

  The ammonia supplies the NH_3 and the ammonium chloride the NH_4^+ ions.

Buffers are a **conjugate pair** formed from a **weak acid + conjugate base** pair or a **weak base + conjugate acid** pair.

- To **calculate acid buffer pH**, use the equation

$$pH = pK_a + \log_{10} \frac{[A^-(aq)]}{[[HA(aq)]} \quad \text{i.e. } pH = pK_a + \log_{10} \frac{[SALT(aq)]}{[ACID(aq)]}$$

Example: Calculate the pH of a buffer solution in which $[CH_3COOH] = 0.1\ mol\,dm^{-3}$ and $[CH_3COO^-Na^+] = 0.05\ mol\,dm^{-3}$ ($CH_3COOH\ pK_a = 4.76$).

$pH = 4.76 + \log_{10} (0.05/0.1) = 4.76 + (-0.30) = 4.46$

Example: Dissolved CO_2 helps to buffer the **blood**. The weak acid is carbonic acid: $CO_2(g) + H_2O(l) \rightleftharpoons H_2CO_3(aq)$
and the conjugate base is the hydrogencarbonate ion $HCO_3^-(aq)$.

Buffers are important in **living systems** because enzymes (biological catalysts) are denatured (stop working) if the pH becomes too extreme.

Example:
Methyl orange:

HIn(aq) (red) ⇌
H^+(aq) + In^-(aq) (yellow)

For **strong alkali**, use **phenolphthalein**.

For **strong acid**, use **methyl orange**.

Use **either** indicator with strong acid – strong base.

- **Indicators** are weak acids (or alkalis) which are coloured. When H^+ is lost (or gained) the colour changes. For an acidic indicator HIn:

  HIn(aq) ⇌ H^+(aq) + In^-(aq)

  The dissociation constant for an indicator is given the symbol K_{ind}.

  $$K_{ind} = \frac{[H^+(aq)][In^-(aq)]}{[HIn]}$$

  You must remember that when there are equal amounts of the two colours, then [HIn(aq)] = [In^-] then K_{ind} = [H^+(aq)], so pK_{ind} = pH. Therefore, the colour changes when **pH = pK_{ind}**. The colour change must occur within the abrupt change of pH. Indicators cannot show a distinct end point in weak acid – weak base titrations.

TESTS

RECALL TEST

1 A Brønsted–Lowry acid is ________________

and a base is ________________ (2)

2 A strong acid is ________________ (1)

3 Give equations for

a pH ________________ (1)

b pK_a ________________ (1)

c pK_w ________________ (1)

d pOH ________________ (1)

4 Find the pH of:

a 0.05 mol dm^{-3} H^+ ions ________________

b 0.05 mol dm^{-3} HNO_3(aq) ________________

c 0.05 mol dm^{-3} H_2SO_4(aq) ________________

d 0.05 mol dm^{-3} CH_3COOH(aq) (when ethanoic acid $K_a = 5 \times 10^{-5}$ mol dm^{-3})

________________ (4)

5 Write an equilibrium equation for the dissocation of methanoic acid (HCOOH).

________________ (1)

6 Find the pH of 0.05 mol dm^{-3} NaOH(aq) (when $K_w = 10^{-14}$ mol^2 dm^{-6}).

________________ (1)

7 What is a buffer solution? ________________ (2)

8 What is the pH of a solution containing 0.05 mol dm^{-3} ethanoic acid and 0.05 mol dm^{-3} sodium ethanoate (when the pK_a of ethanoic acid is 4.76)?

________________ (2)

9 Calculate the pH of a solution of 0.05 mol dm^{-3} ethanoic acid and 0.1 mol dm^{-3} sodium ethanoate.

________________ (2)

10 On another sheet of paper, draw a simple exam-quality titration curve for a strong acid with a weak base. (2)

(Total 20 marks)

CONCEPT TEST

1 a Define

i 'Brønsted–Lowry base' ____________________

ii pH ____________________

iii pK_w ____________________ (3)

b Calculate the pH of a 0.01 mol dm^{-3} sulphuric acid solution.

____

____________________ (2)

c Ethanoic acid has $pK_a = 4.76$, and a concentration of 1.70 mol dm^{-3}. Calculate the pH.

____

____________________ (3)

d At 42 °C K_w is 3.0×10^{-14} $mol^2\ dm^{-6}$. Calculate the pH of pure water at this temperature.

____

____________________ (2)

e At 42 °C, calculate the pH of a 0.001 mol dm^{-3} solution of NaOH.

____

____________________ (2)

2 Carbon dioxide dissolved in blood helps to buffer the blood pH.

$$HCO_3^-(aq) \rightleftharpoons H^+(aq) + CO_3^{2-}(aq)$$

a Use this equation to give the reactions that occur when acid and alkali are added.

acid added ____________________

alkali added ____________________ (2)

b 100 cm^3 of 0.2 mol dm^{-3} $NaHCO_3$ was mixed with 100 cm^3 of 0.1 mol dm^{-3} NaOH.

i Calculate the concentration of $CO_3^{2-}(aq)$ in the mixture.

____

____________________ (2)

ii Calculate the concentration of $HCO_3^-(aq)$ in the mixture.

____

____________________ (2)

iii Calculate the pH of the mixture at 25 °C, given that K_a for the equilibrium above is 4.8×10^{-11} mol dm^{-3}.

____

____________________ (2)

(Total 20 marks)

Unit 19 AROMATIC CHEMISTRY

Aromatic chemistry is the study of compounds based on the **benzene ring** C_6H_6.

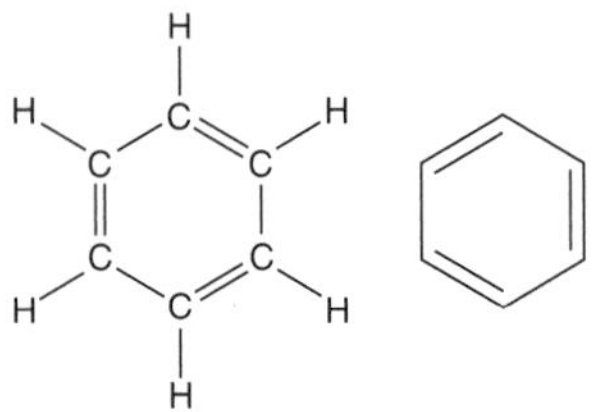

Kekulé structures of benzene (obsolete)

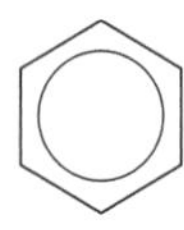

delocalised structure

Fig. 19.1

- Compounds based on the **benzene ring** are called **arenes**. The benzene molecule is a ring of six carbon atoms; each carbon atom is joined to two others and a single hydrogen atom. You will encounter the structure drawn in a number of ways (see Fig. 19.1).

  Benzene is unsaturated, but it does not react with bromine (by addition) like an alkene. It is very unreactive. The six carbon–carbon bonds are of equal length, not alternating single and double bonds (see Fig. 19.2).

Bond	Bond enthalpy ($kJ\ mol^{-1}$)	Bond length (nm)
C–C	348	0.154
C⩦C in benzene	518	0.139
C=C	612	0.134

Fig. 19.2

- If you compare the **hydrogenation enthalpy** for **benzene** with the hydrogenation enthalpy for **three cyclohexene molecules** (3 × C=C), then you will see the hydrogenation of benzene releases less heat (see Fig. 19.3). The **p atomic orbitals** on each carbon atom overlap with each other to form a set of **delocalised molecular orbitals** (see Fig. 19.4). The six delocalised electrons in these orbitals have **lower energy** than the six electrons in 3 C=C bonds. The **difference in the enthalpies** of hydrogenation represents the **extra stability** of the delocalised electron structure.

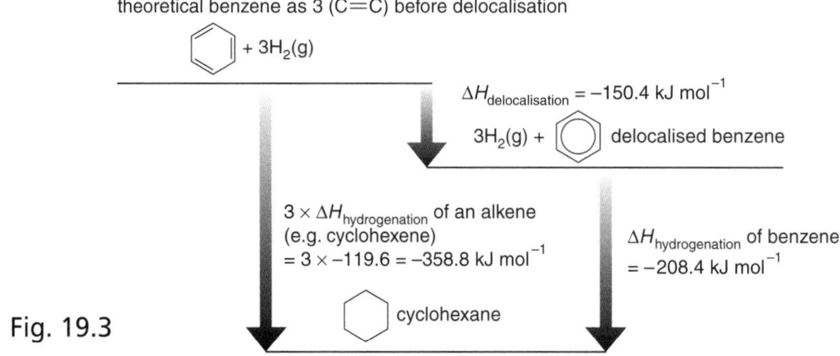

Fig. 19.3

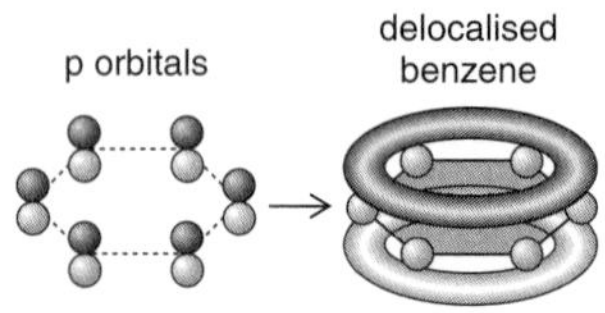

Fig. 19.4

- The stability of the electrons delocalised around the benzene ring means that it does not undergo electrophilic addition. Rather, it undergoes **electrophilic substitution** of the hydrogen atoms by strong electrophiles.

  Typical electrophilic substitution reactions include:

  Nitration to form nitrobenzene: $C_6H_6 + HNO_3$ (in H_2SO_4) $\rightarrow C_6H_5NO_2 + H_2O$

  Concentrated HNO_3 (in H_2SO_4) is the 'nitrating mixture' that provides the nitronium ion electrophile NO_2^+ (see Fig. 19.5).

  Alkylation to form e.g. ethylbenzene (the 'Friedel–Crafts' reaction):

  $C_6H_6 + CH_3CH_2Cl \xrightarrow{AlCl_3} C_6H_5CH_2CH_3 + HCl$ (see Fig. 19.6).

Nitration of benzene

electrophile

$HNO_3 + H_2SO_4 \longrightarrow NO_2^+ + HSO_4^- + H_2O$

NO_2^+ H NO₂ H NO₂ + H^+

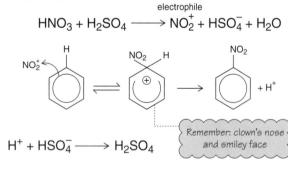

Remember: clown's nose and smiley face

$H^+ + HSO_4^- \longrightarrow H_2SO_4$

Fig. 19.5

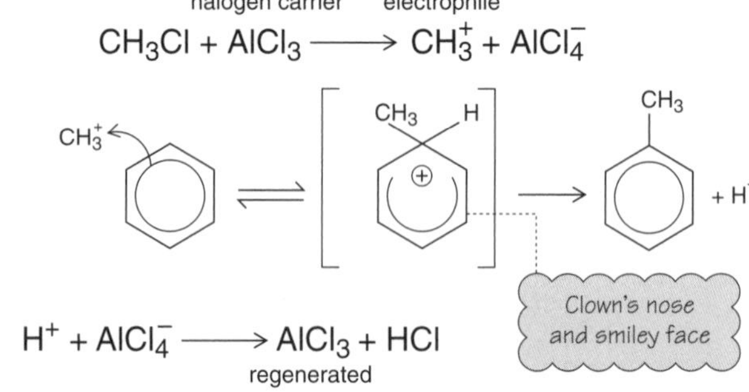

Fig. 19.6

Chlorination to form chlorobenzene: $C_6H_6 + Cl_2 \xrightarrow{AlCl_3} C_6H_5Cl + HCl$

Bromination to form bromobenzene: $C_6H_6 + Br_2 \xrightarrow{AlBr_3} C_6H_5Br + HBr$

Acylation to form e.g. phenylethanone:

$C_6H_6 + CH_3COCl \xrightarrow{AlCl_3} C_6H_5COCH_3 + HCl$

- Note that, except for nitration, **$AlCl_3$** is used as a **catalyst**. It must be used in **anhydrous** conditions: water would hydrolyse it. $AlCl_3$ forms dative (co-ordinate) bonds with the other reagent to produce the positively **charged electrophile** e.g.

  Chlorination: $AlCl_3 + Cl_2 \rightarrow AlCl_4^- + \mathbf{Cl^+}$

  Alkylation: $AlCl_3 + CH_3CH_2Cl \rightarrow AlCl_4^- + \mathbf{CH_3CH_2^+}$

  Acylation: $AlCl_3 + CH_3COCl \rightarrow AlCl_4^- + \mathbf{CH_3CO^+}$

  You can **brominate** benzene with a mixture of **Fe** and **Br_2**. Some of the bromine reacts with the iron to make iron bromide which then acts as the halogen carrier for bromination. See also the manufacture of styrene in unit 16.

- The **$-CH_3$** group on methylbenzene, like all alkyl groups, will be converted into **$-CH_2Cl$** by **homolytic free radical substitution** using chlorine and ultraviolet light (see Fig. 19.7). Contrast this reaction with chlorination of the benzene ring itself, using Cl_2 and $AlCl_3$.

  Surprisingly, the **$-CH_3$** group on **methylbenzene** is easily **oxidised** by aqueous acidified potassium permanganate (manganate(VII)) to make **benzoic acid**. Any other alkyl group (ethyl, propyl, etc.) will also oxidise to benzoic acid.

  $2C_6H_5CH_3 + 7[O] \rightarrow 2C_6H_5COOH + 3H_2O$

- **Phenol** C_6H_5OH is **acidic** due to the p electrons on the O atom being withdrawn into the delocalised ring, allowing H^+ to leave.

  Phenol is a very **weak acid**. It does turn blue litmus to red, but does *not* liberate CO_2 from carbonates. It will react with NaOH(aq).

  $C_6H_5OH(aq) + NaOH(aq) \rightarrow C_6H_5O^-Na^+(aq) + H_2O(l)$

  The phenol group activates the ring to electrophilic substitution, making phenol much **more reactive** than benzene. For example, phenol is nitrated by aqueous nitric acid (see Fig. 19.8) and is brominated by aqueous bromine to make 2,4,6-tribromophenol (TBP), a white suspension (see Fig. 19.9). In a similar way, the antiseptic TCP, 2,4,6-trichlorophenol, is made from phenol and chlorine.

  In common with alcohols, phenol will form esters (see unit 21).

  The **test** for a **phenolic group** is to add aqueous iron III chloride. Any phenolic group will make a particularly evil-looking violet solution (an iron III phenolic complex).

- **Nitrobenzene** may be converted into **phenylamine** by reduction using tin with concentrated hydrochloric acid.

  $C_6H_5NO_2 + 6[H] \rightarrow C_6H_5NH_2 + 2H_2O$

  In common with all amines, phenylamine is basic (see unit 21).

  Phenylamine is converted to a **diazonium ion** using nitrous acid HNO_2 (or HONO). This reagent is unstable so must (i) be kept cold and (ii) be made when needed by mixing a source of nitrite ions (NO_2^-) with H^+ ions, e.g.

  $NaNO_2(aq) + HCl(aq) \rightarrow HNO_2(aq) + NaCl(aq)$

- You must write down the production of **nitrous acid** and the **diazotisation reaction** as shown in Fig. 19.10.

- Ice cold aqueous **phenol** with alkali converts the diazonium ion into an **azo dye** 4-hydroxyphenylazobenzene. Two molecules join together in this **coupling reaction** (see Fig. 19.10).

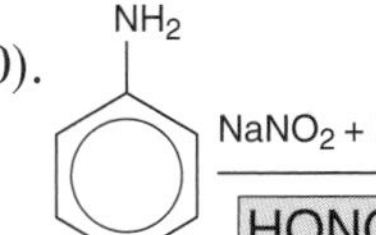

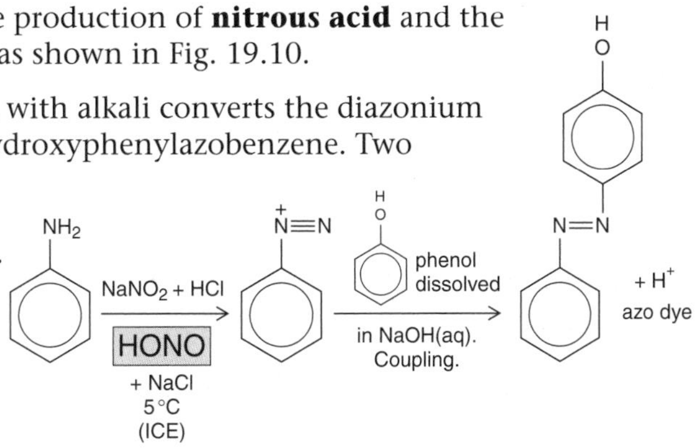

+ NaCl
5°C
(ICE)

Fig. 19.10

$AlCl_3$ is called a **halogen carrier** in halogenation reactions.

Other halogen carriers include $FeCl_3$, $AlBr_3$, and FeBr.

Nitrated aryl compounds can be used as explosives e.g. 'trinitrotoluene' TNT $CH_3C_6H_2(NO_2)_3$.

Nitrobenzene $C_6H_5NO_2$ may be reduced to **phenylamine** $C_6H_5NH_2$ which then converted into azo dyestuffs.

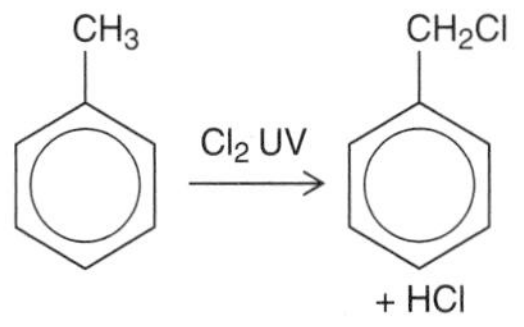

Fig. 19.7

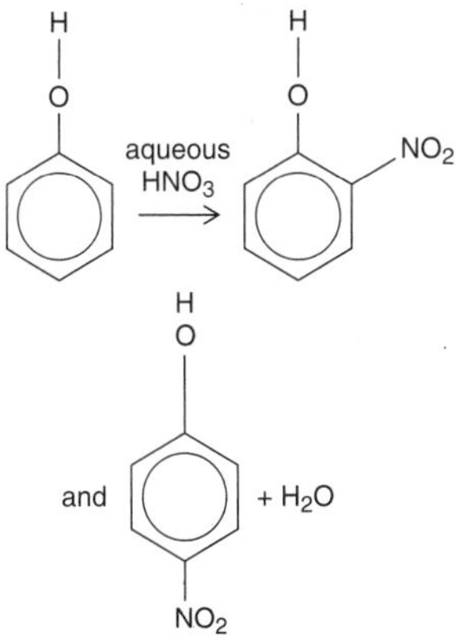

Fig. 19.8

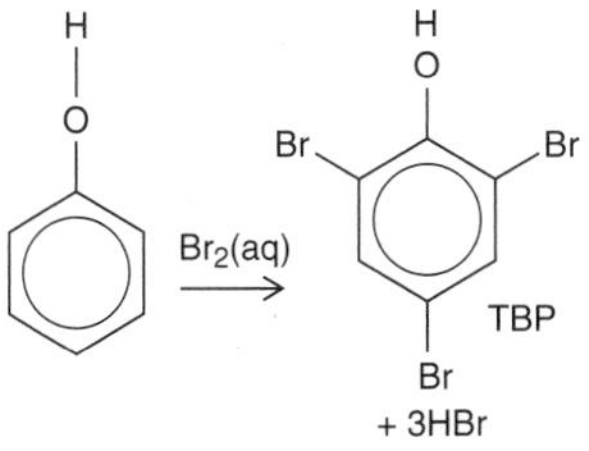

Fig. 19.9

Extensively delocalised structures are often coloured. The molecular structure responsible for the colour is called the **chromophore**. (See also 2,4-DNP.)

Unit 19

TESTS

RECALL TEST

1 Explain why all the carbon–carbon bonds in benzene have the same length, and a length intermediate between the C-C and C=C bond lengths.

(1)

2 Explain how the delocalisation energy may be determined.

(3)

3 On a piece of paper, draw the mechanism for nitration of benzene. (2)

4 On a piece of paper, draw the mechanism for forming methylbenzene from benzene. (2)

5 Write balanced equations for:

a benzene + concentrated nitric acid (with $H_2SO_4(l)$)

b benzene + chlorine (with $AlCl_3$)

c benzene + bromine (with $FeBr_3$)

d benzene + ethanoyl chloride (with $AlCl_3$)

(4)

6 Explain why aluminium chloride must be anhydrous for electrophilic substitution to occur.

(2)

7 How can styrene (phenylethene) be made from benzene?

(2)

8 Give two uses for nitrated aromatic compounds.

(2)

9 State what is made when methylbenzene is oxidised.

(1)

10 Give the reagents and conditions required to convert:

a methylbenzene into benzoic acid

b nitrobenzene into phenylamine

c phenylamine into a diazonium salt

d a diazonium salt into an azo dye

e phenol into sodium phenoxide

f phenol into 2,4,6-trinitrophenol

(6)

(Total 25 marks)

CONCEPT TEST

1 Benzene may be converted into methylbenzene.

a On paper, draw the mechanism for this reaction. (5)

b Name this mechanism. ____________________ (1)

2 Methylbenzene, A, may be converted by a sequence of reactions to produce a very useful compound, E.
First methylbenzene, A, is nitrated using concentrated nitric and sulphuric acids to produce substance B.
Then B is converted into compound C, with the molecular formula C_7H_9N.
C reacts with aqueous sodium nitrite and sulphuric acid, in ice, to form a solution containing the ion D, with the empirical formula $C_7H_7N_2^+$.
Solution D, when mixed with an alkaline phenol solution, in ice, forms a brightly coloured solid, E.
If the solution D were allowed to warm up then a phenol would form.

a Give the structure of B.

____________________ (2)

b Give the reagents required to convert B into C.

____________________ (2)

c Give the structure of the ion D.

____________________ (1)

d What type of substance is E?

____________________ (1)

e On paper, draw the structure of E. (1)

f Give the test for phenol.

____________________ (2)

3 Give the reagents and conditions necessary to carry out the following changes:

a C_6H_6 to $C_6H_5CH_3$

____

b $C_6H_5CH_3$ to $C_6H_5CH_2Cl$

____

c $C_6H_5CH_2Cl$ to $C_6H_5CH_2OH$

____

d $C_6H_5CH_3$ to C_6H_5COOH

____

e C_6H_5OH to $C_6H_5O^-Na^+$

____________________ (5)

(Total 20 marks)

Unit 20 ALDEHYDES, KETONES, AND NITRILES

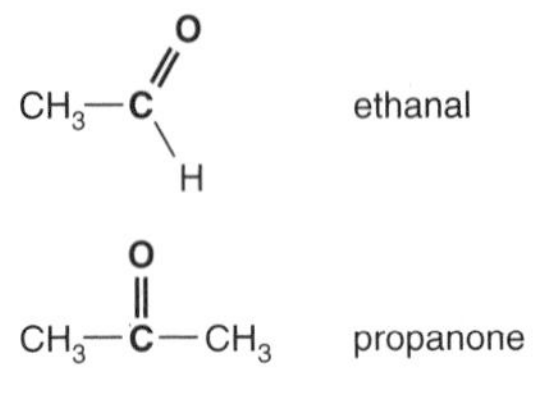

$CH_3—C(=O)—H$ ethanal

$CH_3—C(=O)—CH_3$ propanone

Fig. 20.1

Aldehydes are reduced to primary alcohols.

Ketones are reduced to secondary alcohols.

Fig. 20.2

$CH_3C(=O)H$ + HCN → (pH = 5, KCN (aq)) → $CH_3—C(OH)(H)—C{\equiv}N$ → (heat under reflux, HCl(aq)) → $CH_3—C(OH)(H)—C(=O)OH$

Fig. 20.4

Fig. 20.3

- Aldehydes and ketones are **carbonyl compounds** because they contain the **>C=O** carbonyl group (see Fig. 20.1). The C=O bond is **polarised** and gives each molecule a permanent dipole (but no H-bonding). **Dipole–dipole interactions** give aldehydes and ketones **higher b.p.s** than alkanes of corresponding relative molecular mass.

  Ethanal CH_3CHO (b.p. 21 °C) and **propanone** CH_3COCH_3 (b.p. 56 °C) **burn** in oxygen with a blue flame. **Aldehydes** smell like **apples** and are toxic. Ethanal (made from ethene) is the **starting material** for many organic compounds. As an oxidation product of ethanol, it is responsible for hangovers. **Ketones** have a **sweet smell** and are also toxic. Propanone (acetone), an important **solvent**, was once used as nail varnish remover.

- The **reduction** of aldehydes and ketones gives **alcohols**. $NaBH_4$(aq) or $LiAlH_4$ (dry ether) reduce **aldehydes** to **primary** alcohols and **ketones** to **secondary** alcohols. Use [H] to indicate a reducing agent. **Example:** Ethanal reduced to ethanol. $CH_3CHO + 2[H] \rightarrow CH_3CH_2OH$

  Example: Propanone reduction: $CH_3COCH_3 + 2[H] \rightarrow CH_3CH(OH)CH_3$

  Reduction occurs because the **H^- ion** acts as a **nucleophile** (see Fig. 20.2). Note that H^+ ions are also needed, supplied either by water or by the later addition of aqueous acid.

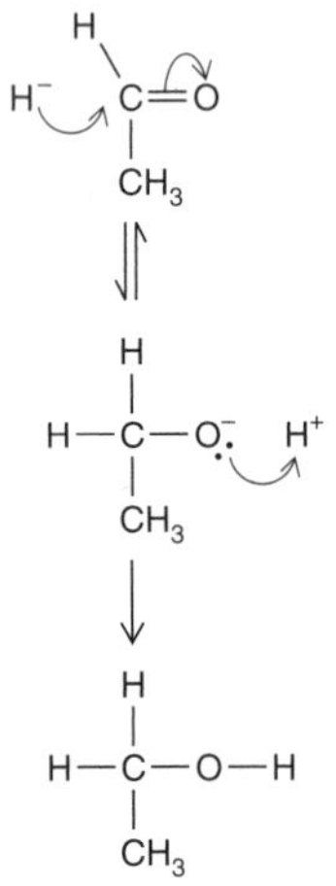

- Most common **oxidising agents** (e.g. acidified aqueous dichromate) will readily oxidise **aldehydes** to **carboxylic acids**. Ketones resist oxidation. **Example:** $CH_3CHO + [O] \rightarrow CH_3COOH$

- Use Fehling's solution or Tollens' reagent to **distinguish** between **aldehydes** and **ketones**. Only aldehydes are oxidised by these reagents and give a positive result.

  Fehling's solution contains **Cu^{2+}** ions which are **reduced** to a red solid, copper I oxide (**Cu_2O**).

  $2Cu^{2+}(aq) + H_2O(l) + 2e^- \rightarrow Cu_2O(s) + 2H^+(aq)$

  Tollen's reagent ('ammoniacal silver nitrate') contains **Ag^+** ions which are reduced during oxidation of the aldehyde to a distinctive **silver mirror** on the inside of the test tube (or else a black precipitate of silver).

  $Ag^+(NH_3(aq)) + e^- \rightarrow Ag(s)$

  To make **Tollen's reagent**: (i) Add aqueous **NaOH** to aqueous **silver nitrate** to make a pale brown precipitate of silver hydroxide; (ii) drip aqueous **ammonia** into the solution until the precipitate disappears. To **test** for an aldehyde, add the organic substance to this solution and warm gently.

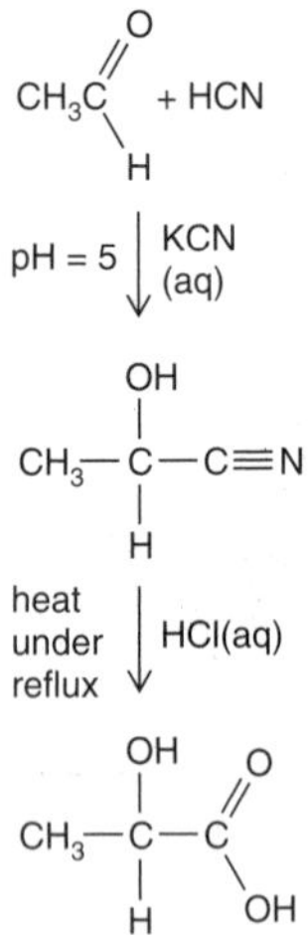

- Aldehydes and ketones react by **nucleophilic addition**, i.e. heterolytic nucleophilic addition occurs when HX reacts with any carbonyl compound.

  $CH_3COCH_3 + HX \rightarrow CH_3COHXCH_3$

  See Fig. 20.3 for the mechanism of **HCN addition**. Note that the **nucleophile** is the **CN^-** ion; the **attacking lone pair** is on the carbon atom **:CN^-**. The pH must be only **slightly acidic** or else the reaction is too slow. The optimum pH is 5. If the **pH is too low**, the many H^+ ions join the CN^- ions to make HCN, lowering $[CN^-]$ and **slowing the first step** of the reaction. If the **pH is too high** then $[H^+]$ is low, **slowing the second step**.

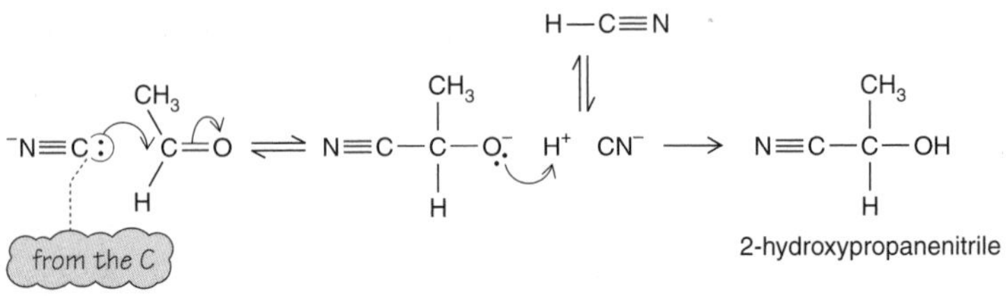

- **Addition** of **HCN** to the C=O group produces **two functional groups**: a hydroxy group **-OH** and a nitrile **-CN**. In examination questions, the nitrile group is often **hydrolysed** to make a **carboxylic acid** (see Fig. 20.4).

- To **test** for aldehydes and ketones, you add **2,4-dinitrophenylhydrazine** (2,4-DNP) (see Fig. 20.5) to make **dinitrophenylhydrazones**, which are brightly coloured **yellow-orange solids**. 2,4-dinitrophenylhydrazine reacts by a **condensation** reaction. **Example: Ethanal** to make **ethanal 2,4-dinitrophenylhydrazone** (see Fig. 20.6).

  You must know the full name of 2,4-dinitrophenylhydrazine and know its structure (see Fig. 20.5). It helps to remember that **hydrazine** is NH_2NH_2 and that **phenyl** indicates a benzene ring; then add two **nitro groups** ($-NO_2$) at positions **2** and **4** on the ring. You must also be able to draw the structures of dinitrophenylhydrazones.

  Each dinitrophenylhydrazone, purified by recrystallization, has a **distinctive melting point** that indicates which carbonyl compound was originally present (see Fig. 20.7 for melting point apparatus).

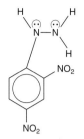

2,4-dinitrophenylhydrazine

Fig. 20.5

ethanal 2,4-dinitrophenylhydrazone

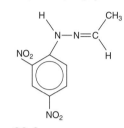

Fig. 20.6

> **Test for:** $CH_3-C(=O)-$ or $CH_3-CH(OH)-$
>
> **Note:** A CH_3- next to the C=O
> **Reagents:** iodine in aqueous alkali (or aqueous KI in NaClO)
> **Positive result:** Pale yellow precipitate
> **Products: Iodoform (CHI_3) and a salt of carboxylic acid**
> **Example:** CH_3COCH_3 with I_2 in NaOH(aq) produces CHI_3 and $CH_3COO^-Na^+$

Fig. 20.8

> The **iodoform (triiodomethane) test** is interesting as it gives information on the **structure** of a molecule rather than just identifying the functional group present (see Fig. 20.8).

- Nitriles are **made** either (i) by reacting a **halogenoalkane** with **cyanide ion** e.g. $CH_3Br + HCN \rightarrow CH_3CN + HBr$

  or by dehydration of an amide using P_4O_{10} e.g. $CH_3CONH_2 \rightarrow CH_3CN + H_2O$

- **Nitriles** can be converted to **amines** by **reduction**, using hydrogen gas and a platinum (or nickel) catalyst, or $NaBH_4$(aq), or $LiAlH_4$(dry ether).
  Example: Propanenitrile reduces to propylamine.

  $CH_3CH_2CN + 4[H] \rightarrow CH_3CH_2CH_2NH_2$

- Acid or alkaline **hydrolysis** will form **carboxylic acids** from nitriles.
  Example: Heating ethanenitrile CH_3CN under reflux with aqueous **HCl** produces **ethanoic acid** CH_3COOH. However, hydrolysis of CH_3CN by aqueous **NaOH** produces **sodium ethanoate** $CH_3CO_2^-Na^+$.

- Grignard reagents are **organometallic compounds** with the general formula **RMgX**, where X is a halogen (usually Br).

  Grignard reagents are **prepared** by adding **magnesium** metal to a **halogenoalkane** in dry ether, with free halogen as a catalyst.
  Example: Ethylmagnesium bromide.

  $CH_3CH_2Br + Mg \xrightarrow{Br_2} CH_3CH_2MgBr$

  Grignard reagents react with **water** to form an **alkane**.
  Example: $CH_3CH_2MgBr + H_2O \rightarrow CH_3CH_3 + Mg(OH)Br(aq)$

  Grignard reagents **attack C=O** groups by **nucleophilic addition**; the final product forms when H^+(aq) is added.

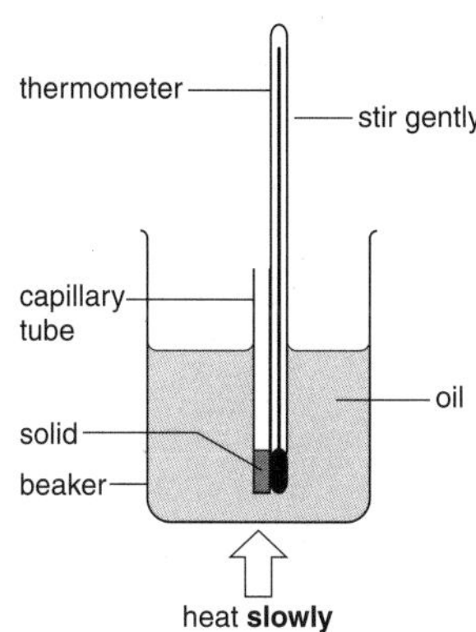

Fig. 20.7 Melting point apparatus

> methanal HCHO $\rightarrow$ **primary alcohols**
> $CH_3CH_2MgBr + HCHO \rightarrow CH_3CH_2CH_2OH + Mg(OH)Br$
>
> aldehydes RCHO $\rightarrow$ **secondary alcohols**
> $CH_3CHO + CH_3CH_2MgBr \rightarrow CH_3CH_2(CH_3)CHOH + Mg(OH)Br$
>
> ketones RCOR' $\rightarrow$ **tertiary alcohols**
> $CH_3COCH_3 + RMgBr \rightarrow (CH_3)_2RCOH + Mg(OH)Br$
>
> Carbon dioxide $\rightarrow$ **carboxylic acids**
> $CO_2 + RMgBr \rightarrow RCOOH + Mg(OH)Br$

> These notes about **Grignard reagents** are required by Edexcel/London Board candidates only.

> **Nitriles** have the **-C≡N** functional group (see for example Fig. 20.9).

$CH_3-CH_2-C\equiv N$

propanenitrile

Fig. 20.9

Grignard reagents **attack acyl chlorides** by **nucleophilic substitution** to form **ketones**. **Example:**

$CH_3CH_2MgBr + CH_3COCl \rightarrow CH_3CH_2COCH_3$ (plus $Mg^{2+}/Br^-/Cl^-$ ions)

Unit 20

TESTS

RECALL TEST

1 What intermolecular forces do aldehydes and ketones have?

____________________________________ (1)

2 Both aldehydes and ketones often react by the mechanism ____________________. Both may be reduced to form an ____________________. Only ____________________ may be oxidised to a ____________________. (5)

3 On a sheet of paper, draw the mechanism for the reaction between propanone and HCN. (3)

4 In the reaction between propanone and HCN the pH is critical.

a If the pH is too low, why is the reaction too slow?

____________________________________ (1)

b If the pH is too high, why is the reaction too slow?

____________________________________ (1)

5 Name the product of the reaction between ethanal and HCN.

____________________________________ (1)

6 State the test which is positive for both ketones and aldehydes:

Reagent ____________________________________

What happens ____________________________________ (2)

7 **a** On a piece of paper, draw 2,4-dinitrophenylhydrazine. (1)

b On a piece of paper, draw the product of the reaction between 2,4-dinitrophenylhydrazine and ethanal, and name it. (2)

8 State the reaction type and the product formed when $NaBH_4$(aq) is reacted with:

a propanal ____________________________________

b propanone ____________________________________ (2)

9 Write a balanced equation for these reactions.

a $NaBH_4$(aq) with propanal ____________________________________

b $LiAlH_4$(dry ether) with propanone ____________________________________

c HCN with ethanal ____________________________________

d ethanal with an oxidising agent ____________________________________ (4)

10 Name the two tests that are positive with aldehydes but not ketones.

____________________________________ (2)

11 Propanone may be converted into 2-hydroxy-2-methylpropanenitrile (A), and that then converted into 2-hydroxy-2-methylpropanoic acid (B). State how the conversions may be carried out:

Propanone to A ____________________________________

A to B ____________________________________ (2)

12 Give the reagents and conditions for these conversions:

a ethanenitrile to ethanoic acid ____________________________________

b ethanenitrile to ethylamine ____________________________________

c bromomethane to ethanenitrile ____________________________________ (3)

(Total 30 marks)

CONCEPT TEST

1 Compound X has the molecular formula C_3H_6O. X reacts with 2,4-dinitrophenylhydrazine and with a solution of aqueous iodine and sodium hydroxide (when it forms a pale yellow solid).
X may be converted by two steps into compound Z, $C_4H_8O_3$, via a compound Y, C_4H_7NO.

Step I X $\rightarrow$ Y

Step II Y $\rightarrow$ Z

a Use the information above to draw the structure of X on a piece of paper. (1)

b Step I occurs by nucleophilic addition. Draw this mechanism. (3)

c If the pH of step I is too alkaline then the rate will be very slow. Explain why this is so.

______________________________ (2)

d Give the reagents and conditions for step II.

______________________________ (2)

e Draw the structure of 2,4-dinitrophenylhydrazine. (1)

f Give a test that will distinguish between an aldehyde and a ketone.

______________________________ (2)

g How may the 2,4-dinitrophenylhydrazine derivative be purified?

______________________________ (1)

h Explain how the pure derivative may be used to confirm the identity of X.

______________________________ (2)

2 Aqueous sodium boron tetrahydride ($NaBH_4$) reacts with aldehydes, ketones, and nitriles differently.

a What type of reagent is $NaBH_4$?

______________________________ (1)

b Give the structural formula of the product when $NaBH_4(aq)$ is reacted with:

i propanol ______________________________

ii propanone ______________________________

iii propanenitrile ______________________________ (3)

c Give a balanced equation for the reaction of:

i $NaBH_4(aq)$ with ethanal ______________________________

ii $KMnO_4(aq)$ with ethanal ______________________________ (2)

(Total 20 marks)

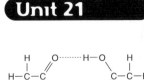

Unit 21 CARBOXYLIC ACIDS AND AMINES

Fig. 21.1 Ethanoic acid dimerises

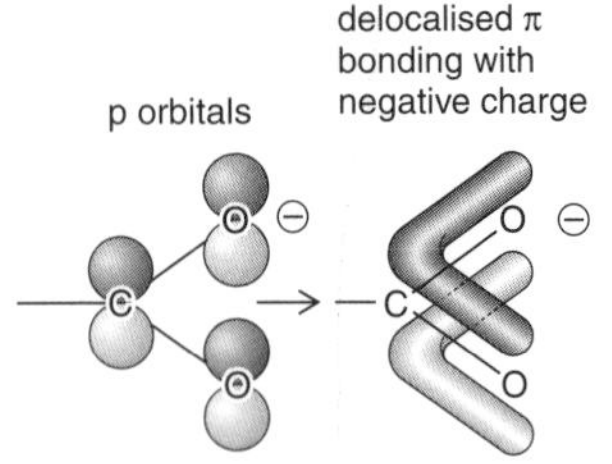

Fig. 21.2

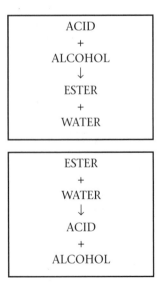

The **carboxylic acid group** is written as **COOH** (not CO_2H) to show that the oxygen atoms are different.

The **carboxylate ion** is written $\mathbf{CO_2^-}$ (not COO^-) to show that the two oxygen atoms are equivalent.

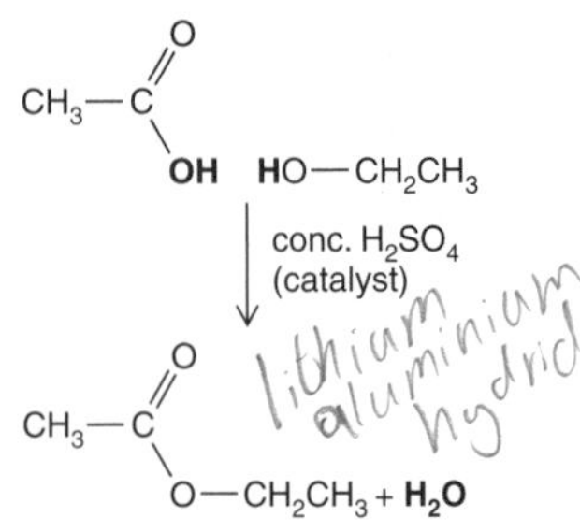

Fig. 21.3

ACID
+
ALCOHOL
↓
ESTER
+
WATER

ESTER
+
WATER
↓
ACID
+
ALCOHOL

- **Carboxylic acids** contain the functional group **-COOH** (see Fig. 21.1). The C=O and the O-H bonds are **polar**, giving extensive **H-bonding** and high m.p.s and b.p.s. When pure liquids, or in non-polar solvents, carboxylic acids **dimerise**. Two molecules H-bond to form a single **non-polar dimer**.

  Carboxylic acids form stable $\mathbf{-CO_2^-}$ ions because of **delocalisation** of the negative charge around the three atomic centres. The C and O atoms all have similar sizes which allows the p orbitals to overlap easily (see Fig. 21.2).

- Carboxylic acids **dissociate incompletely** in aqueous solution and so are **weakly acidic**. The sour taste of vinegar that improves the flavour of fish and chips is due to the formation of $\mathbf{H^+(aq)}$ ions.

  $CH_3COOH(aq) \rightleftharpoons CH_3CO_2^-(aq) + H^+(aq)$

  Remember that the alternative representation is:

  $CH_3COOH(aq) + H_2O(l) \rightleftharpoons CH_3CO_2^-(aq) + H_3O^+(aq)$

  The **strength** of a carboxylic acid depends on how easily H^+ can leave, which depends on the **electron density** of the O-H group. If there is an electron-donating group (e.g. $-CH_3$), causing a positive inductive effect (+I), then the O-H bond is electron rich; it is more difficult for H^+ to leave and the acid is **weaker**. If the group is **electron withdrawing** (e.g. -Cl or a benzene ring), giving a negative inductive effect (–I), then the O-H bond is electron deficient; it is easier for H^+ to leave and the acid is **stronger**. The inductive effect explains the increase in acidity: $CH_3COOH < C_6H_5COOH < HCOOH$.

- **Carboxylic acids** react as typical acids. They turn litmus from **blue to red** and **neutralise bases**. **Example:**

  $CH_3COOH(aq) + NaOH(aq) \rightarrow CH_3CO_2^-Na^+(aq) + H_2O(l)$

  They liberate **carbon dioxide** (with effervescence) from carbonates and hydrogencarbonates. **Example:**

  $CH_3COOH(aq) + NaHCO_3(s) \rightarrow CH_3CO_2^-Na^+(aq) + H_2O(l) + CO_2(g)$

- Carboxylic acids react with **alcohols** to make **esters** and water (see Fig. 21.3). **Example:**

  $CH_3COOH(aq) + CH_3CH_2OH(aq) \rightleftharpoons CH_3COOCH_2CH_3(aq) + H_2O(l)$

- Carboxylic acids can be **reduced** to **alcohols**. The powerful reducing agent $LiAlH_4$ (in dry ether) **must** be used. **Example:**

  $CHCOOH(aq) + 4[H] \rightarrow CH_3CH_2OH(l) + H_2O(aq)$

- Many esters have **pleasant smells**. Fruity ethyl ethanoate is used as a food flavouring (e.g. in sweets) and as an ink solvent in some 'spirit' felt marker pens. Some esters are used in perfumes, as solvents, or as plasticisers to soften plastics. Esters are **hydrolysed** by heating under reflux with **dilute aqueous acid** (e.g. H_2SO_4) to form a **carboxylic acid** and an **alcohol**. (The $H^+(aq)$ ions catalyse the reaction.) **Example:**

  $CH_3COOCH_2CH_3(l) + H_2O(l) \rightleftharpoons CH_3COOH(l) + CH_3CH_2OH(l)$

  Esters are also **hydrolysed** by heating under reflux with dilute alkali (aqueous base) to form the **salt** of the carboxylic acid and an **alcohol**. (The $OH^-(aq)$ ions catalyse the reaction.) **Example:**

  $CH_3COOCH_2CH_3(l) + NaOH(aq) \rightarrow CH_3CO_2^-Na^+(aq) + CH_3CH_2OH(aq)$

  Fats and **oils** are **triglyceride esters** (esters of the trihydric alcohol, glycerol $CH_2OHCHOHCH_2OH$). Their alkaline hydrolysis (which is called **saponification**) is used to manufacture **soap**.

- $\mathbf{PCl_5(s)}$ will substitute -Cl for the -OH group on a carboxylic acid, to form an **acyl chloride** (also called acid chloride or carbonyl chloride) -COCl. The reaction is vigorous and produces white fumes of HCl. Example (to form ethanoyl chloride):

  $CH_3COOH(l) + PCl_5(s) \rightarrow CH_3COCl(l) + POCl_3(s) + HCl(g)$

- **Acyl chlorides** (see Fig 21.4) react like **very reactive carboxylic acids**. The C-Cl bond is **easily broken** because the Cl atom is too large to form a pi bond with the C atom. Acyl chlorides undergo **condensation** (nucleophilic addition–elimination) reactions at room temperature in which the **Cl** atom is **eliminated** (see Fig. 21.5).

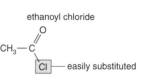

ethanoyl chloride

$CH_3—C(=O)—Cl$ — easily substituted

Fig. 21.4

- Acyl chlorides react with the **nucleophiles** water (forming **carboxylic acids**), alcohols (forming **esters**), ammonia (forming **amides**), and amines. **Examples:**

  With water: $CH_3COCl(l) + H_2O(l) \rightarrow CH_3COOH(aq) + HCl(g)$

  With alcohol: $CH_3COCl(l) + CH_3CH_2OH(l) \rightarrow CH_3COOCH_2CH_3(aq) + HCl(g)$

  With ammonia: $CH_3COCl(l) + 2NH_3(g) \rightarrow CH_3CONH_2(s) + NH_4Cl(s)$

  With amines: $CH_3COCl(l) + CH_3CH_2NH_2(g) \rightarrow CH_3CONHCH_2CH_3(s) + HCl(g)$

  The -CONH- group is called an **amide link**. In proteins, it is called a **peptide link**.

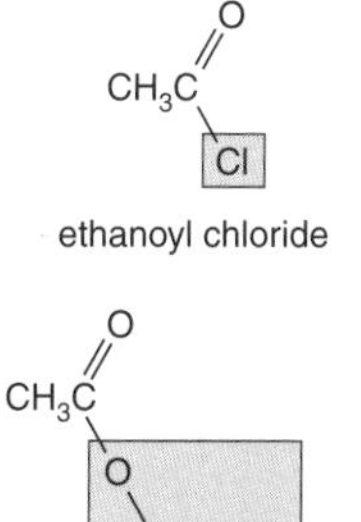

Fig. 21.5

- In common with alcohols, phenol will form esters. However, a condensation reaction using a reactive acyl chloride is the best choice, e.g.

  $C_6H_5OH + CH_3COCl \rightarrow CH_3COOC_6H_5 + HCl$

  The **acid anhydrides** react in a similar way to acyl chlorides.

  $(CH_3CO)_2O + HOC_6H_4COOH \rightarrow CH_3COOC_6H_4COOH$

- **Aspirin** is an ester made from 2-hydroxybenzoic acid and ethanoic anhydride. The by-product is ethanoic acid. Ethanoyl chloride is not used because the strongly acidic HCl would be left in the aspirin.

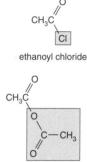

With ethanol,
acyl chlorides are very reactive
(use room temperature),
acid anhydrides are less reactive
(a little heat helps),
carboxylic acids are the least reactive
(heat under reflux).

$RCOCl > (RCO)_2O > RCOOH$

- Amines contain the **$-NH_2$** functional group. They can act as proton acceptors (bases) and as **nucleophiles**. The polar N-H group enable amines to **H-bond**, so they have high m.p.s and b.p.s.

- The $-NH_2$ lone pair can attract a proton (H^+ ion) and bond to it:

  $R\text{-}NH_2 + H^+ \rightarrow R\text{-}NH_3^+$

  The **attractive power** of the lone pair determines the $-NH_2$ **base strength**. If the $-NH_2$ group is joined to an **electron-donating group** (the +I inductive effect of e.g. $-CH_3$ or $-CH_2CH_3$, etc.) then the lone pair is electron rich and so the $-NH_2$ group is **more basic**. If the $-NH_2$ group is joined to an **electron-withdrawing group** (the –I inductive effect of e.g. the delocalised electrons in an aromatic ring) then the lone pair is electron deficient and the $-NH_2$ group is **less basic**. The inductive effect explains the increase in basicity.

  Amines are basic so they react with **acids** to make **salts**. **Example:** ethylamine forms the salt ethyl ammonium chloride.

  $CH_3CH_2NH_2 + HCl \rightarrow CH_3CH_2NH_3^+Cl^-$

$CH_3CH_2NH_2$

This is aminoethane, also called ethylamine.

- In common with all **amines**, phenylamine is **basic**. It has limited solubility in water but **dissolves** readily in acid to form the aqueous **phenylammonium ion**.

  $C_6H_5NH_2(l) + H^+(aq) \rightarrow C_6H_5NH_3^+(aq)$

  which will react with alkali to **regenerate** the free base.

  $C_6HNH_3^+(aq) + OH^-(aq) \rightarrow C_6H_5NH_2(l) + H_2O(l)$

Ethylamine is more basic than ammonia is more basic than phenylamine.

- Ammonia will **react repeatedly** with a **halogenoalkane** by nucleophilic substitution to make various amines. **Example:**

  (i) $NH_3 + CH_3Br \rightarrow HBr + CH_3NH_2$ (a **primary** amine)

  (ii) $CH_3NH_2 + CH_3Br \rightarrow HBr + (CH_3)_2NH$ (a **secondary** amine)

  (iii) $(CH_3)_2NH + CH_3Br \rightarrow HBr + (CH_3)_3N$ (a **tertiary** amine)

  (iv) $(CH_3)_3N + CH_3Br \rightleftharpoons Br^- + (CH_3)_4N^+$ (a **quaternary** ammonium ion)

  Long-chain quaternary ammonium ions are used as **detergents** (cationic surfactants) in bubble bath liquids and washing powders. They are also **mildly antiseptic** and are used in throat lozenges (look on ingredients labels).

Primary aliphatic amines are best prepared by the **reduction** of **nitriles**, using $NaBH_4(aq)$. **Aromatic amines** are prepared by **reducing aromatic nitro** compounds using tin and concentrated HCl.

TESTS

RECALL TEST

1 Write balanced equations for the reaction between ethanoic acid and:

a NaOH(aq) __________

b $NaHCO_3$(aq) __________

c PCl_5(s) __________

d ethanol (with conc. H_2SO_4) __________

e $LiAlH_4$ (dry ether) __________ **(5)**

2 State four tests for carboxylic acids:

____

____

__________ **(4)**

3 State the reagents and conditions required to make ethyl propanoate.

__________ **(2)**

4 Write a balanced equation for the acid hydrolysis of ethyl ethanoate.

__________ **(1)**

5 State how soap may be made from fat.

__________ **(1)**

6 Write balanced equations for the following reactions:

a $CH_3COCl + H_2O \rightarrow$ __________

b $CH_3COCl + CH_3CH_2OH \rightarrow$ __________

c $CH_3COCl + NH_3 \rightarrow$ __________

d $CH_3COCl + CH_3CH_2NH_2 \rightarrow$ __________

e $CH_3NH_2 + HCl \rightarrow$ __________

f $NH_3 + CH_3Br \rightarrow$ __________

g $CH_3NH_2 + CH_3Br \rightarrow$ __________ **(7)**

7 State the reagents and conditions required to convert the first substance into the second:

a CH_3COCl into CH_3COOCH_3 __________

b CH_3COCl into $CH_3CONHCH_3$ __________

c $CH_3NH_3^+$ into CH_3NH_2 __________

d CH_3COOH into CH_3COCl __________ **(8)**

8 Give the reagents and conditions required to split these molecules:

a $CH_3CH_2OCOCH_3$ __________

b $CH_3CH_2NHCOCH_3$ __________ **(2)**

(Total 30 marks)

CONCEPT TEST

1 This question concerns the following reactions:
$C_2H_4O_2$ (P) is converted by PCl_5 to C_2H_3ClO (Q).
Q is converted by addition of ethanol to $C_4H_8O_2$ (R).
R is converted by aqueous NaOH to $C_2H_3O_2Na$ (S) and C_2H_6O (T).
P also may be converted by $LiAlH_4$(aq) to T.
P reacts with NaOH(aq) to produce S.
T reacts with P to produce R.
CH_5N (U) reacts with Q to produce C_3H_7NO (V).

a Give the structures of P to V:

P ____________________

Q ____________________

R ____________________

S ____________________

T ____________________

U ____________________

V ____________________ **(7)**

b Using structural formulae, write equations for:

i the conversion of P to S

____

ii the conversion of Q to R

____

iii the conversion of U and Q to V

____

iv the conversion of R to S

____________________ **(4)**

c State the conditions required to react:

i Q with T to make R ____________________

ii P to make S ____________________

iii R to make S and T ____________________ **(6)**

2 In this question you are given the name of the compounds and you must identify the reagents and conditions or draw the mechanism on a sheet of paper.

a Aminomethane reacts with bromomethane. Draw the mechanism for this reaction. **(4)**

b Aminomethane reacts with ethanoyl chloride to make $CH_3CONHCH_3$.

i State the conditions required. ____________________

ii Name this compound, $CH_3CONHCH_3$. ____________________

iii Name the type of compound. ____________________ **(3)**

c Give a test, and the expected result, that would distinguish between:

i CH_3COOH and CH_3COCl. ____________________

ii CH_3COOCH_3 and CH_3COOH. ____________________ **(6)**

(Total 30 marks)

Unit 22

POLYMERS AND OPTICAL ISOMERS

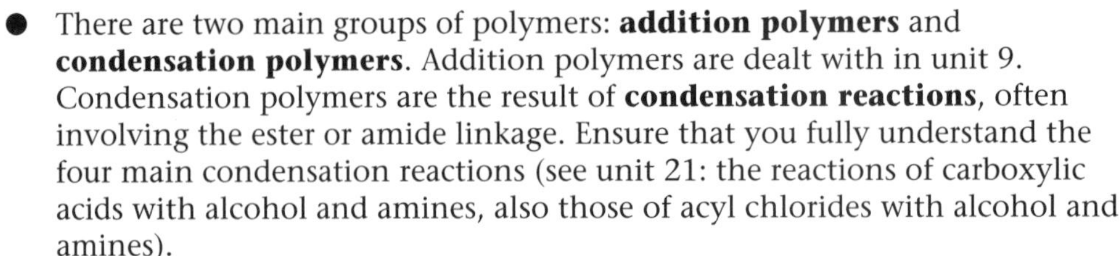

- There are two main groups of polymers: **addition polymers** and **condensation polymers**. Addition polymers are dealt with in unit 9. Condensation polymers are the result of **condensation reactions**, often involving the ester or amide linkage. Ensure that you fully understand the four main condensation reactions (see unit 21: the reactions of carboxylic acids with alcohol and amines, also those of acyl chlorides with alcohol and amines).

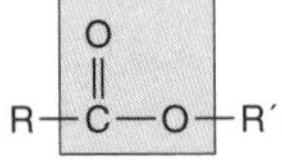

Fig. 22.1 Ester link

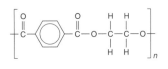

Fig. 22.2 Terylene repeating unit

- **Polyesters** are produced when **dicarboxylic acid** molecules **HOOC**-[R]-**COOH** condense with **diol** (dihydric alcohol) molecules **HO**-[R']-**OH**. R and R' are alkyl or aryl hydrocarbon residues. **Ester links** form between alternating acid and alcohol molecules (see Fig. 22.1). For example, benzene 1,4-dicarboxylic acid $HOOC(C_6H)COOH$ condenses with ethane-1,2-diol $HOCH_2CH_2OH$ to form **Terylene**, a common synthetic polyester fibre used in cloth and rope manufacture (see Fig. 22.2).

  Separate polyester chains are held together by **dipole–dipole** attraction due to the permanent **polarisation** in the **ester** groups, so polyesters have higher softening points than polyalkanes. These polymers are not crystalline, so they do not have sharp melting points.

- You must know one **biodegradable** polymer, such as **Biopol**. It is made from a single monomer **3-hydroxybutanoic acid** $HOCH(CH_3)CH_2COOH$, which polymerises by forming ester links between the -OH and -COOH groups. The repeating unit is $[\text{-}OCH(CH_3)CH_2COO\text{-}]_n$. Bacteria use this substance as an energy store. Potatoes could be **genetically engineered** to produce this monomer instead of starch in their tubers.

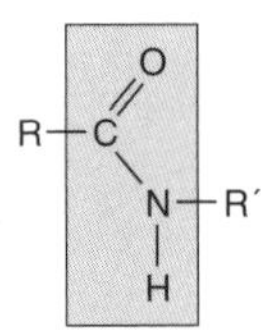

Fig. 22.3 Amide link

- **Polyamides** such as **Nylon** are manufactured from a **diamine** $H_2N[R]NH_2$ and a **dicarboxylic acid** HOOC-[R']-COOH. For example, **Nylon 6,6** is made from 1,6-diaminohexane $H_2N(CH_2)_6NH_2$ and hexane-1,6-dioic acid $HOOC(CH_2)_4COOH$. Each monomer (made from benzene C_6H_6) contains 6 C atoms, hence the name Nylon 6,6. The repeating polymer unit is $[\text{-}HN(CH_2)_6NHCO(CH_2)_4COO\text{-}]_n$. **Amide links** couple the monomer units together (see Fig. 22.3).

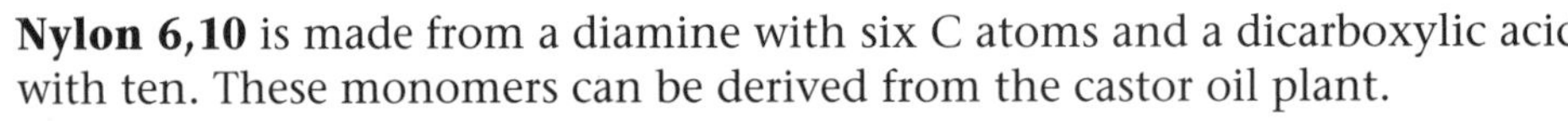

  Nylon 6,10 is made from a diamine with six C atoms and a dicarboxylic acid with ten. These monomers can be derived from the castor oil plant.

  A stronger type of Nylon is made by incorporating a benzene ring $\text{-}C_6H_4\text{-}$ into the structure. An example is **Kevlar**, which is used to make bullet-proof vests and puncture-proof canoes and tyres.

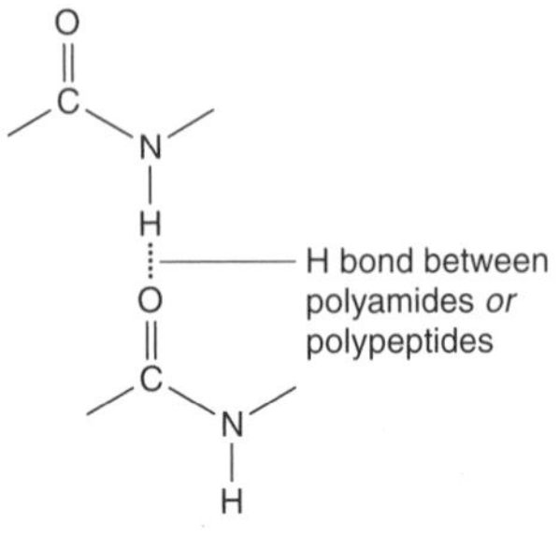

Fig. 22.4

  The Nylon polymer chains are strongly attracted to each other by **hydrogen** bonding between the -NH groups and the >C=O groups (see Fig. 22.4). Nylon is used as a fibre in clothes and carpets and in the heavy-duty ropes used to tie up ships. This tough plastic material is also used to make **hard-wearing** mechanical parts such as washing machine valves and food mixer gears.

- Because of their manufacture by **condensation reactions**, it should not surprise you to know that both polyesters and polyamides will be **hydrolysed** by prolonged exposure to aqueous acids and alkalis.

- You must be able to recognise a polymer's **type** (addition, condensation, polyester, polyamide, etc.) from either the **repeating unit** (general formula) or a sample of the **chain**. Look for the type of link (amide, ester, or simple C-C) and then suggest the **monomers** used to form the links.

  When given monomer **names** or **structures**, you must be able to deduce the **type** of polymer and give the **repeating unit**.

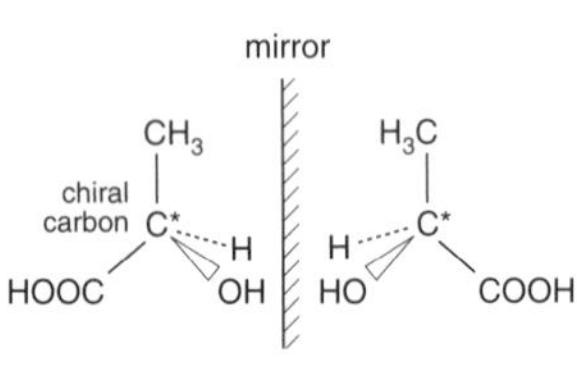

Fig. 22.5 Lactic acid shows optical isomerism.

- **Optical isomers** rotate the plane of **polarised light** in opposite directions. Two optical isomers (also called **enantiomers**) are mirror images of each other and cannot be superimposed on each other (see Fig. 22.5). Optical isomerism arises when a carbon atom (called a **chiral centre**) has four different atoms or groups attached to it.

Note that you must draw the isomers clearly in **three dimensions**. Take care to make the shape **tetrahedral** about the chiral centre.

- **Amino acids** are substances that have both an **amine group** -NH_2 and a **carboxylic group** -COOH (see Fig. 22.6). All naturally occurring amino acids have the general formula $H_2NCHRCOOH$. R may be a H atom or any one of a large variety of side chains. If R is not H then the C in the amino acid is **chiral**, so most amino acids have optical isomers.

  In the solid state and in aqueous solution, the acidic -COOH group donates a proton (H^+ ion) to the -NH_2 group, forming a double ion that is called a **zwitterion** (see Fig. 22.7).

  $H_2NCHRCOOH \rightleftharpoons {}^+H_3NCHRCOO^-$

  This ion has acidic and basic properties.

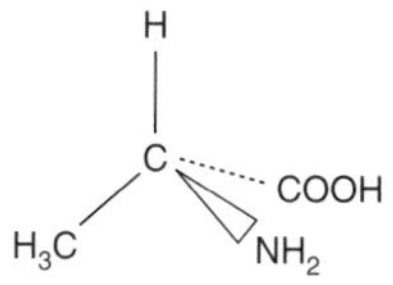

Fig. 22.6 Alanine, a simple amino acid

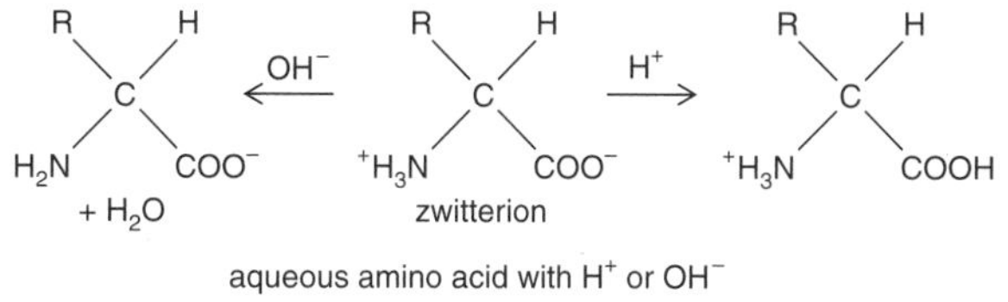

aqueous amino acid with H^+ or OH^-

Fig. 22.7

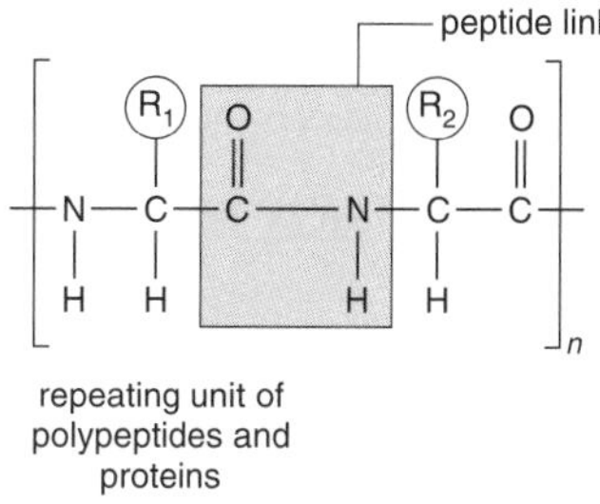

repeating unit of polypeptides and proteins

Fig. 22.8

- Amino acids condense together in chains to form **polypeptides**. The amino acids are joined by **peptide links**, which are the same as the **amide link** (see Fig. 22.8). Polypeptides join together to make **proteins**.

- 2-hydroxypropanoic acid (**lactic acid**) is found **naturally** in milk. It forms in your muscles when you run hard and makes them ache. Natural lactic acid always consists of just one optical isomer of 2-hydroxypropanoic acid because **enzymes** are **stereospecific** and only make one of the enantiomers.

  Lactic acid can be **synthesised** by reacting ethanal with HCN (at pH 5) to make 2-hydroxypropanenitrile. (Note the use of cyanide to add one C atom to the molecule.) Heating the product with aqueous acid under reflux forms 2-hydroxypropanoic acid. The product is a **50–50 mixture** of the two optical isomers and is called a **racemic mixture**. The isomers have an equal and opposite effect on polarised light and so **do not rotate** its plane of polarisation.

- Optical activity has **medical** implications. If a drug has a chiral centre, one isomer may be beneficial and the other may be **inactive** or even **harmful**. Manufacturers must ensure their products are pure and contain only the beneficial enantiomer. **Example: Ibuprofen**, a pain killer and anti-inflammatory, has a chiral centre (see Fig. 22.9)

  The purity of drugs is often improved by **recrystallisation**. Thin layer chromatography (**TLC**) is used as a simple check on purity and composition.

  A **drug** is any substance that has a physiological effect. Potential new drugs are isolated by medical research. Each is called an NCE (**new chemical entity**). If it has a damaging effect on life it is called a **poison**. If it has a beneficial effect it is called a **medicine**. Once a medicine has been proved safe and is given to patients without a doctor's prescription it is called an OTC (**over the counter drug**). Some medicines are harmful if given to the wrong patients. For example, aspirin must not be given to children or people with stomach ulcers. Drugs have a specific effect due to their **shape**. Hydrogen-bonding side groups such as -OH, NH_2, and -COOH help to make these substance **water-soluble**.

  Much research is carried out to develop more effective **antibiotics**. One great problem is the growth of **antibiotic-resistant** bacterial strains, brought about by the **indiscriminate use** of antibiotics and patients not completing courses of treatment.

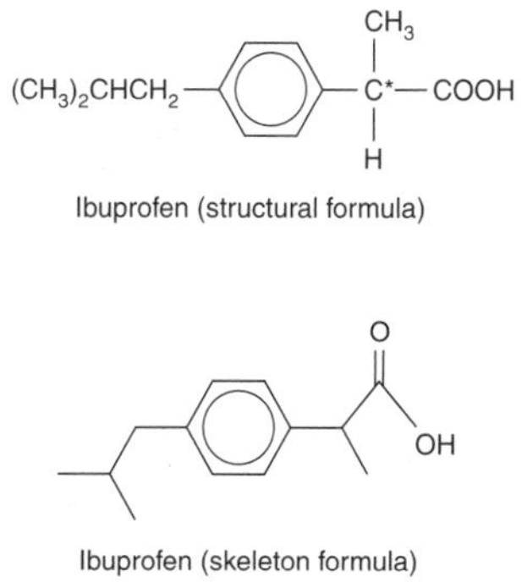

Ibuprofen (structural formula)

Ibuprofen (skeleton formula)

Fig. 22.9

Unit 22

TESTS

RECALL TEST

1 State the specific monomers required to make Terylene, a polyester.

________________________________ (2)

2 On a separate sheet, draw the repeating unit of

a A polyester
b Nylon 6,10
c Polyhydroxybutanoic acid (PHB)
d A polypeptide. (4)

3 Give an example of a substance used to make cross links between the polyester chains in Terylene.

________________________________ (1)

4 State the monomers required to make:

a Nylon 6,6 ________________

b Kevlar ________________

c protein ________________ (3)

5 Give a large-scale use for each of these polymers:

a Terylene ________________

b Nylon 6,6 ________________

c Kevlar ________________ (3)

6 Boiling polyester and polyamides in aqueous alkali would ____________ the long-chain molecules. (1)

7 Give the formula of the species when:

a amino acids are dissolved in water ________________

b acid is added to this solution ________________

c alkali is added to the original solution ________________ (3)

8 State the type of polymer in each case (addition, polyester, polyamide, polypeptide):

a $[\text{-CH}_2\text{-CH(CH}_3)\text{-}]_n$ ________________

b $[\text{-NHCH}_2\text{CONHCH}_2\text{CONHCH(CH}_3)\text{CO-}]_n$ ________________

c $[\text{-O(CH}_2)_2\text{OCO(C}_6\text{H}_4)\text{CO-}]_n$ ________________

d $[\text{-CO(C}_6\text{H}_4)\text{CONH(C}_6\text{H}_4)\text{NH-}]_n$ ________________

e $[\text{-OCH(CH}_3)\text{CH}_2\text{CO-}]\text{n}$ ________________ (5)

9 State what is meant by 'optical isomerism'.

____

________________________________ (2)

10 Indicate which of the carbon atoms are chiral in this molecule:

$\text{HOCH}_2\text{CHClCH}_2\text{CH(CH}_3)\text{COOH}$ (2)

11 Explain how optical isomers are distinguished.

________________________________ (2)

12 On a sheet of paper, draw the optical isomers of lactic acid (2-hydroxypropanoic acid). (2)

(Total 30 marks)

CONCEPT TEST

1 Examine Fig. 22.10 below, which shows fragments of polymers.

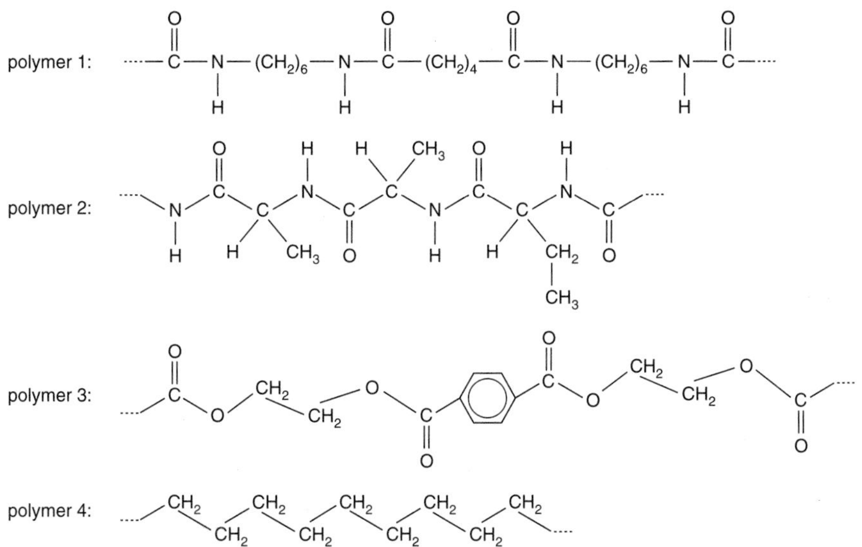

Fig. 22.10

a Give the structural formula of the monomer(s) required for each polymer:

Polymer 1 ____________________

Polymer 2 ____________________

Polymer 3 ____________________

Polymer 4 ____________________ (7)

b State the type of polymer in each case:

Polymer 1 ____________________

Polymer 2 ____________________

Polymer 3 ____________________

Polymer 4 ____________________ (4)

c State the intermolecular forces in each case:

Polymer 1 ____________________

Polymer 2 ____________________

Polymer 3 ____________________

Polymer 4 ____________________ (2)

2 a Give the repeating unit for each polymer produced by these monomers:

i $ClOC(C_6H_4)COCl$ with $H_2N(C_6H_4)NH_2$

ii $H_2N(CH_2)_5COOH$

iii $HO(CH_2)_2OH$ with $ClOC(C_6H_4)COCl$ (6)

b Give the type of the reaction required to make the polymer in part **a i**.

____________________ (1)

(Total 20 marks)

Unit 23 ANALYSIS AND SPECTROSCOPY

You must use all the evidence provided and not try to guess the answer from one clue.

You may need other evidence before you can suggest a structural formula (so be patient).

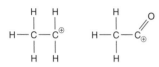

- You are likely to meet questions in examinations that combine organic **analysis**, mass **spectroscopy**, infrared spectroscopy, and nuclear magnetic resonance (NMR) spectroscopy. Your aim is to use the evidence to **identify** an 'unknown compound' or suggest its structural formula.
- **Logical thinking** and **clear written presentation** will sort out which functional groups are present. You may also be able to calculate the relative formula mass (M_r).
- **Mass spectra** are discussed in unit 2. Remember that the **molecular ion** peak on the right of a spectrum gives the **relative molecular mass** of the compound. Other peaks will tell you the masses of the **fragments**.

The more **stable** the fragment, the **higher** the corresponding peak on the spectrum. Carbonium ions (**carbocations** containing a positively charged C atom) and the acylium (RCO^+) ion are both relatively stable (see Fig. 23.1).

Fig. 23.1 These fragments are relatively stable.

Example: One morning a student was found semi-conscious. A blood sample showed the presence of a low molecular mass compound. A sample separated by **gas chromatography** was fed into a mass spectrometer. The spectrum showed that the unknown compound had a **M_r of 46**, with fragments at **15 (CH_3^+)** and **31 ($^+CH_2OH$)**. Forensic scientists suspected that the substance was **ethanol**. A simple chemical test confirmed their suspicions.

Molecular ions tend to break at a **branch** in the carbon skeleton to form carbonium ions, rather than lose individual atoms.

- **Chemical tests** are much more specific, so are often carried out when there is some **clear idea** about the nature of the compound.

Common reagents	Positive test	Functional groups possible
Bromine solution	turns red/brown to colourless	alkene (>C=C<) or phenol (oily drops)
PCl_5(s)	vigorous reaction: white fumes of HCl(g)	-OH group (alcohol, carboxylic acid, water)
2,4-dinitro-phenylhydrazine	brightly coloured solid	ketone or aldehyde
Fehling's solution	red solid forms	aldehyde present
Tollens' reagent	silver mirror or black solid	aldehyde present
$KHCO_3$(s)	fizzes: evolves CO_2	acid present (-COOH but not phenol)
Iodoform test (I_2 in alkali or KI with NaClO)	pale yellow solid	$-COCH_3$ or $-CH(OH)CH_3$

For each test you must know the **reagent**, any **observations**, and what a **positive result** indicates. Often one test will not tell you the functional group, but two together will.

To test for a **halogenoalkane**: heat the substance under reflux with NaOH(aq) to **hydrolyse**, neutralise with nitric acid, and add aqueous **silver nitrate**. Halogenoalkanes will produce a **precipitate**. White indicates -Cl, off-white -Br, pale yellow -I (see unit 10).

Aldehydes are tested for using **Tollen's** reagent or **Fehling's** solution (see unit 20).

- To decide whether an **alcohol** is primary, secondary, or tertiary: either (i) add **$ZnCl_2$** in concentrated HCl (immediate cloudiness indicates a **tertiary** alcohol; cloudiness within 5 minutes indicates a **secondary** alcohol; no change within 5 minutes indicates a **primary** alcohol) or (ii) add acidified **potassium dichromate**; colour change from orange to green indicates a **primary** or a **secondary** alcohol. Then test for an **aldehyde**; a positive result indicates a **primary** alcohol.

One ester smells of muscle rub: methyl 2-hydroxybenzoate ('Oil of Wintergreen').

- **Esters** generally smell **fruity**. Chemical **analysis** involves **hydrolysis** followed by identification of the **carboxylic acid** and the **alcohol**.

Remember that energy is proportional to frequency.

- **Infrared spectroscopy** depends on bonds absorbing infrared radiation to increase their vibrational energy. Different bond types absorb at different frequencies. The **spectrum** shows percentage transmittance of energy (on the *y*-axis) against the frequency of the radiation (on the *x*-axis). The unit of frequency is the **wavenumber** cm^{-1} (waves per centimetre), chosen to give a convenient scale.

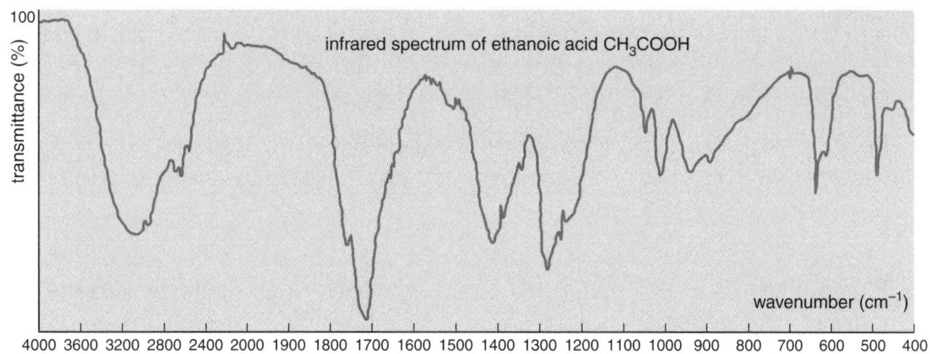

Example: (see Fig. 23.2) The infrared spectrum of **ethanoic acid** CH_3COOH shows **absorbances** at both 1720 cm^{-1} and 2950 cm^{-1} which indicate a carboxylic acid.

Fig. 23.2

- NMR spectra give information about the **positions** of atoms in a molecule. The protons must be **unpaired** so hydrogen protons are usually investigated. Each hydrogen proton in a molecule absorbs energy from **radio waves** at a particular frequency called the **resonant frequency**. Protons in different **environments** (positions) absorb at slightly different frequencies. Tetramethylsilane (**TMS**) $(CH_3)_4Si$ is used as a reference to calibrate NMR equipment because it gives a single peak on the spectrum. The methyl group hydrogen atoms all have the **same environment**. The TMS absorption is labelled **zero** and all other absorbancies are measured with reference to it.

  A simple **low-resolution** NMR spectrum shows vertical absorptions along the *x*-axis which is calibrated as a **chemical shift** in frequency relative to TMS = 0. You compare the spectrum against a table of chemical shifts for different proton environments (see Fig. 23.6).

You don't need to explain how a nuclear magnetic resonance (**NMR**) spectrometer works.

Any solvent must be **hydrogen-free** so solvents such as **tetrachloromethane** are used.

The **area** of each peak reflects the **relative number of protons** in that environment. **Example:** Ethanol CH_3CH_2OH has 3 H atoms in CH_3, 2 H atoms in CH_2, and 1 H atom in OH. Therefore the relative areas of the peaks CH_3:CH_2:OH will be 3:2:1 (see Figs 23.3 and 23.4).

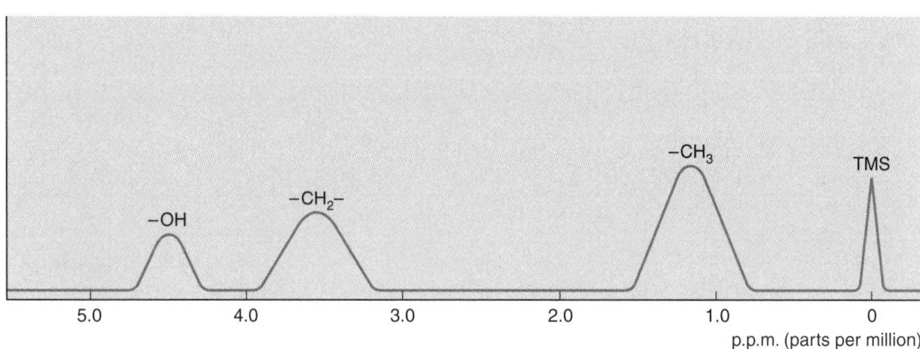

Fig. 23.3

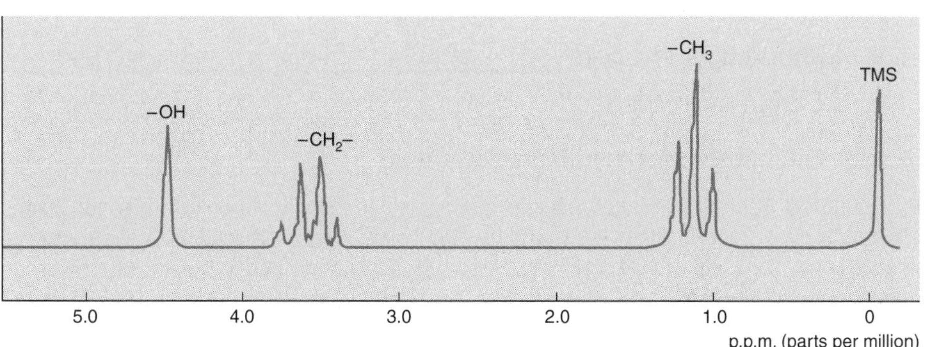

Fig. 23.4

A **high-resolution** NMR spectrum shows that each peak is actually made of little peaks due to magnetic interference (**spin–spin splitting**) from nearby protons.

In the high-resolution ethanol spectrum (see Figs 23.3 and 23.4), the CH_3 peak is split into 3 smaller peaks. The number of little peaks is equal to 1 plus the number of H atoms joined to the next atom. In this case the CH_2 group **splits** the CH_3 peak into 1 + 2 = 3 smaller peaks. The CH_3 group splits the CH_2 peak into 1 + 3 = 4 smaller peaks.

- **Ultraviolet spectroscopy** helps to identify molecules by probing **electron transitions** between energy levels in molecular orbitals. **Visible spectra** usually refer to **transition metal ions** in solution. Remember that the colour we see is due to absorption of frequencies from white light. The **amount** of visible light energy absorbed depends on the **concentration** of the coloured solution.

$[Cu(H_2O)_6]^{2+}$ i.e. $Cu^{2+}(aq)$ is blue because only blue light is **transmitted**, while other frequencies are **absorbed**.

Unit 23

Bond	Wave number (cm^{-1})
C-H (alkane)	2850–2960
C-H (alkene)	3010–3900
C=C	1640–1680
C-O	1150–1310
C=O	1700–1740
O-H (hydrogen bonded)	3200–3550
O-H (not hydrogen bonded)	3590–3650
O-H (carboxylic acid)	2500–3300

Fig. 23.5 Infrared absorption wave numbers

Type of proton	Chemical shift δ/p.p.m.
RCH_3	0.9
R_2CH_2	1.3
R_3CH	1.5
$R_2C{=}CH_2$	–5
C_6H_5–H	7.3
C_6H_5–CH_3	2.3
R–C(=O)–CH_3	2.3
R–C(=O)–H	9.7

Fig. 23.6 Table of approximate chemical shifts and corresponding proton environments. Here R is an alkyl group.

TESTS

RECALL TEST

1 Give the tests for each of these functional groups (the marks indicate the number of tests. Remember to state what occurs when the test is positive):

a alkene

(2)

b alcohol

(4)

c carboxylic acid

(4)

d ketone or aldehyde

(1)

e aldehyde only

(2)

f the group CH_3CO-

(1)

g a bromoalkane

(1)

h phenol

(2)

2 Burning a sample in the fume cupboard may indicate something about a substance. State how the following burn differently:

a a short-chain alcohol (1)

b an aromatic compound (1)

3 Molecules absorb infrared light because of (1)

(Total 20 marks)

CONCEPT TEST

1 A mixture of oils were found in an ancient pottery container. A small sample was separated using steam distillation and gas chromatography. Three smelly components were P, cinnamaldehyde, $C_6H_5CHCHCHO$ (which smells of cimmamon), Q, Oil of Wintergreen, $HOC_6H_4COOCH_3$ (which smells of muscle rub), and R, eugenol, $CH_2CHCH_2C_6H_3(OH)OCH_3$ (which smells of incense).

a State which compound(s) would react with the following reagents. State what would be observed and which functional group is detected.

i Bromine water

(3)

ii Aqueous iron(III) chloride

(3)

iii Aqueous potassium dichromate acidified with sulphuric acid

(3)

b See Fig. 23.7, the infrared spectrum of one of the compounds. Identify the compound and state the evidence that supports your answer. (3)

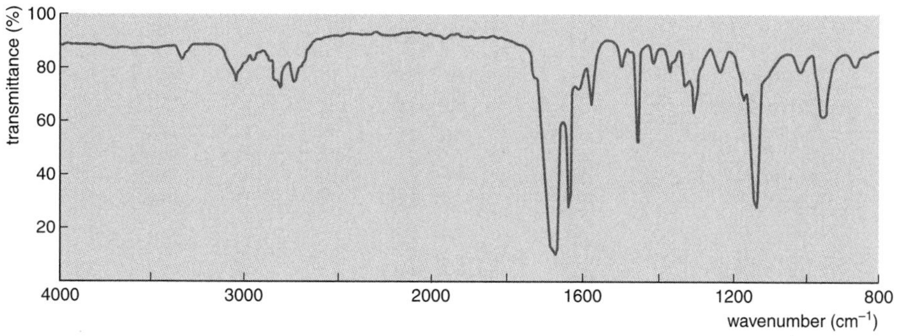

Fig. 23.7

2 Compound X had a mass spectrum which produced peaks with mass/charge ratios of 17, 29, 76, 93, 105, and 122. The tallest peak was 76 and the shortest was 29. The following compounds could be X: CH_3COCH_2OH, CH_3CH_2COOH, $HOCH_2CH_2CH_3$, HOC_6H_4CHO, C_6H_5COOH, or $C_6H_5COCH_3$.

a Suggest the relative molecular mass of X.

(1)

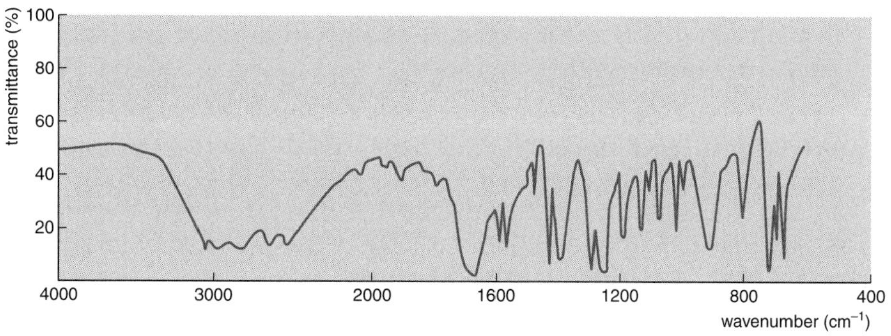

Fig. 23.8

b See Fig. 23.8, which shows the infrared spectrum of X, and, using the mass spectrum data, suggest what functional group(s) may be present in X, stating any evidence that supports your answer.

(4)

c Identify X.

(1)

d Give the chemical shift(s) you would expect for 4-hydroxybenzaldehyde, HOC_6H_4CHO.

(2)

(Total 20 marks)

Unit 24

TRANSITION METALS: PHYSICAL

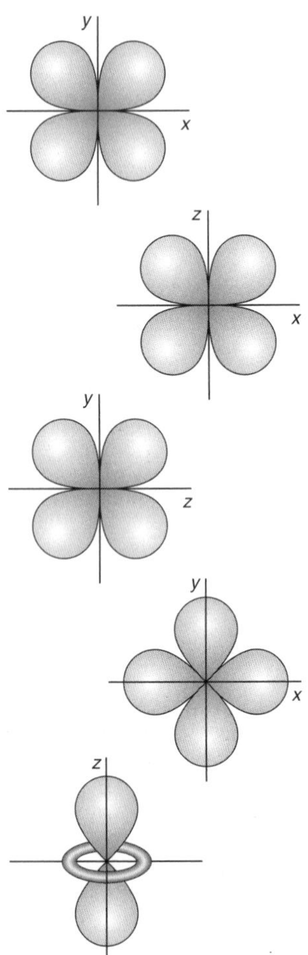

Fig. 24.1

- The **d-block** elements occupy the **central part** of the periodic table in periods 4, 5, and 6. A set of **5 d orbitals** fill with a total of 10 electrons in the course of each period. You must concentrate on the metals of period 4 from **scandium** to **zinc**, in which the 3d orbitals fill from Sc = $3d^1\ 4s^2$ to Zn = $3d^{10}\ 4s^2$. Note that the 4s fills **before** the 3d.

  Four of the d orbitals are shaped like a four-leaf clover and one is shaped like a p orbital inside a doughnut ring (see Fig. 24.1).

- All the elements from Sc to Zn have an **inner shell** electronic configuration $\mathbf{1s^2\ 2s^2\ 2p^6\ 3s^2\ 3p^6}$. These elements are listed in the table below:

d-block element	Symbol	Outer shell electronic configuration	Most important oxidation numbers
Scandium	Sc	$3d^1\ 4s^2$	+3
Titanium	Ti	$3d^2\ 4s^2$	+2 +3 +4
Vanadium	V	$3d^3\ 4s^2$	+2 +3 +4 +5
Chromium	Cr	$\mathbf{3d^5\ 4s^1}$	+2 +3 +6
Manganese	Mn	$3d^5\ 4s^2$	+2 +4 +7
Iron	Fe	$3d^6\ 4s^2$	+2 +3
Cobalt	Co	$3d^7\ 4s^2$	+2 +3
Nickel	Ni	$3d^8\ 4s^2$	+2
Copper	Cu	$\mathbf{3d^{10}\ 4s^1}$	+1 +2
Zinc	Zn	$3d^{10}\ 4s^2$	+2

Note that the 4s has **higher energy** than the 3d. When ions form, the **4s** electrons are lost **before** the **3d** electrons. For example, Fe = $3d^6\ 4s^2$; Fe^{2+} = $3d^6\ 4s^0$ and Fe^{3+} = $3d^5\ 4s^0$. Also, the $\mathbf{d^5}$ and the $\mathbf{d^{10}}$ configurations have **extra stability**, hence Cr = $3d^5\ 4s^1$ rather than $3d^4\ 4s^2$ and Cu = $3d^{10}\ 4s^1$ rather than $3d^9\ 4s^2$. Similarly, Fe^{2+} ($3d^6\ 4s^0$) oxidises easily to Fe^{3+} ($3d^5\ 4s^0$) but it is difficult to oxidise Mn^{2+} ($3d^5\ 4s^0$) to Mn^{3+} ($3d^4\ 4s^0$) because it means breaking into the more stable $3d^5$ configuration.

The table also shows the common oxidation numbers, which you must know. Many elements **first** lose **all 4s** electrons to form an ion. For the elements up to Mn, the **maximum oxidation number** is equal to the **electron number** in the outer shell.

- A **transition element** is an element that forms at least one ion with a **partially filled d subshell**. They have **variable oxidation number** and the aquated ions are **coloured**. Sc has a single oxidation number only, Sc^{3+} = $\mathbf{3d^0}\ 4s^0$. The empty d subshell means that Sc is **not** a transition element. Zn has a single oxidation number only, Zn^{2+} = $\mathbf{3d^{10}}\ 4s^0$. The full d subshell means that Zn is **not** a transition element. $Sc^{3+}(aq)$ and $Zn^{2+}(aq)$ are **colourless**.

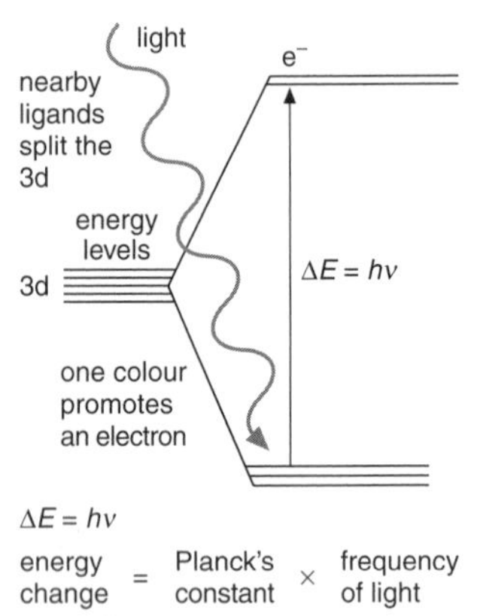

Fig. 24.2

- For an aqueous ion to be **coloured**, the 3d sublevel must be **partially filled**. Nearby **ligands** (see below) **split** the d subshell energy levels. Certain colours (frequencies) are **absorbed** by electrons which are **promoted** from their ground-state d orbital to a higher d orbital. The difference between the energy levels is equivalent to visible (and ultraviolet) parts of the spectrum. If violet/UV is absorbed, then the rest of the colours are transmitted, producing a lemon colour (see Fig. 24.2).

  Colour is determined by the element, the oxidation number, and the ligands (and sometimes the co-ordination number and shape).

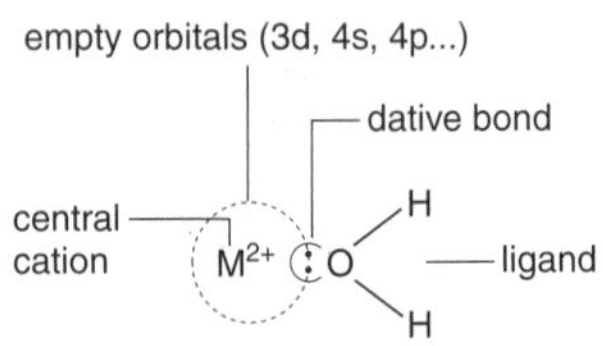

Fig. 24.3 A complex

- **Complexes** form when a central cation (or atom) forms **dative covalent** (co-ordinate) bonds by **accepting** electron pairs from ions (or molecules) called **ligands** (see Fig. 24.3). Ligands donate electrons from lone pairs or pi bonds.

- Ligands that use **one** electron pair per molecule in complexes are called **monodentate** or unidentate ('one-toothed') ligands. Polydentate (multidentate) ligands donate two or more pairs: **bidentate** ligands donate **2** pairs; **tetradentate 4** pairs; **hexadentate 6** pairs. The ligands you must know are listed in Figs 24.4, 24.5, 24.6.

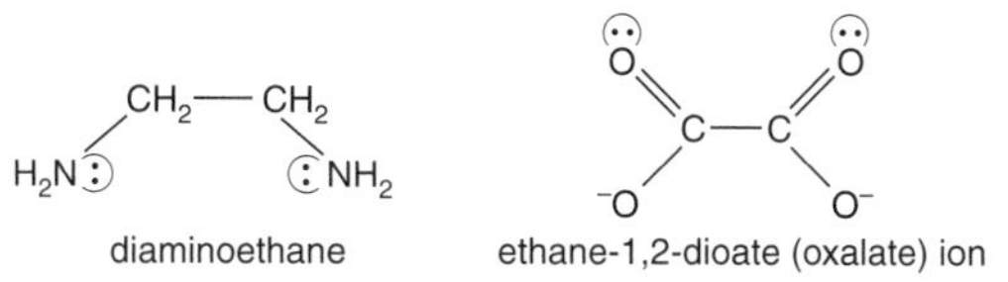

diaminoethane ethane-1,2-dioate (oxalate) ion

Fig. 24.5

Formula	Name
:NH_3	ammine
H_2O:	aqua
:OH^-	hydroxo
:O^{2-}	oxo
:Cl^-	chloro
:CN^-	cyano
:SCN^-	thiocyano
:CO	carbonyl

Fig. 24.4

- The hexadentate ligand **EDTA** is used in shampoo, plant food, and some vitamins because it binds with free aqueous ions. It is also used to remove aquated ions from the bodies of people suffering from lead and cadmium poisoning.

EDTA spread out

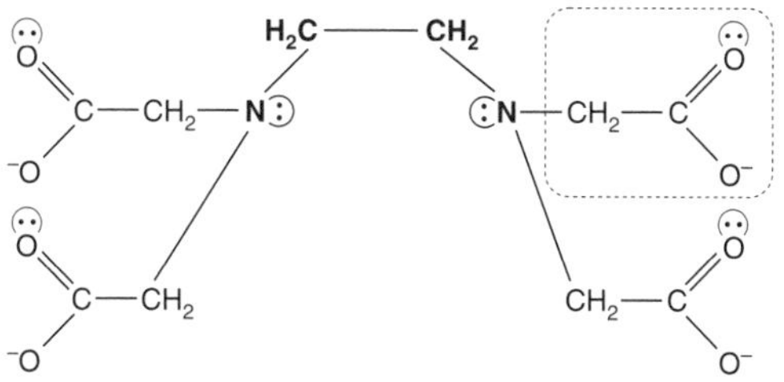

Fig. 24.6 EDTA

Complexes may have 6, 4, or 2 **electron pairs** binding with the **central ion** or **atom**. The number of electron pairs involved is called the **co-ordination number**. Co-ordination number 6 is the most common, 4 is less common, and 2 you will only meet with Ag^+ and Cu^+ ions.

AQA only: a molecule or ion that donates a lone pair is called a **Lewis base**. One that accepts a lone pair is called a **Lewis acid**.

Complexes may be cationic, neutral, or anionic. You must know the shape of complexes. **6**-co-ordinate complexes are **octahedral**, **4**-co-ordinate are usually **tetrahedral** with large ligands (occasionally **square planar** e.g. Ni^{2+}) and **2**-co-ordinate are always **linear**. **Examples:** $[Cr(H_2O)_3(OH)_3]$ is octahedral and neutral; $[CuCl_4]^{2-}$ is tetrahedral and anionic; $[Ag(NH_3)_2]^+$ is linear and cationic. $Ni(CO)_4$ contains a neutral atom and 4 CO molecules.

- Complexes can show **isomerism**. **Example: Geometric isomerism** is shown by $Ni(NH_3)_2Cl_2$; the Cl atoms may be next to (cis) or opposite each other (trans) (see Fig. 24.7). **Example: Optical isomerism** is shown by any complex accepting six lone pairs from 3 bidentate ligands e.g. $[Ni(NH_2CH_2CH_2NH_2]^{2+}$ (see Fig. 24.8).

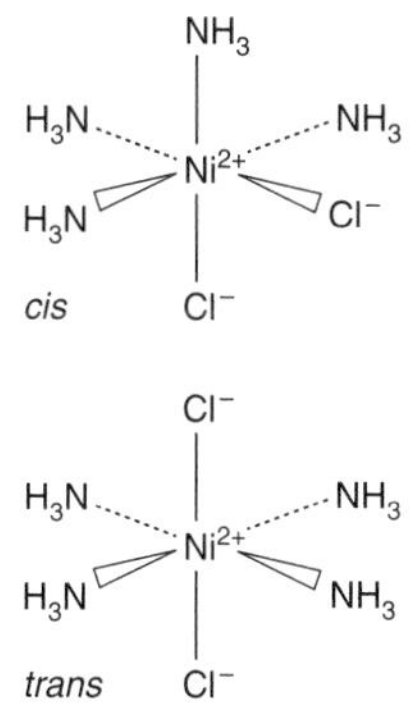

cis *trans*

Fig. 24.7

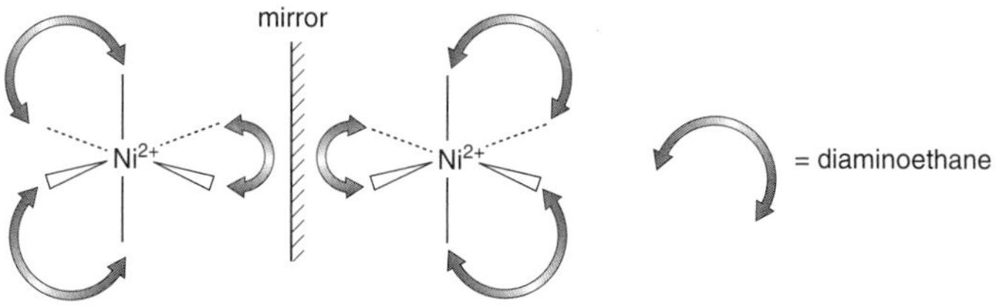

mirror = diaminoethane

Fig. 24.8

The Pt(II) complex ***cis*-platin** is used to treat cancer, while the *trans* form is not effective (see Fig. 24.9).

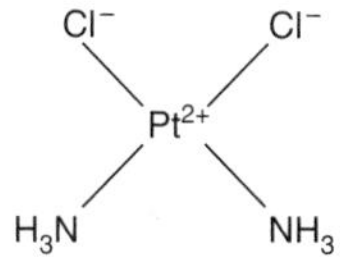

cis-platin is an anti-cancer drug

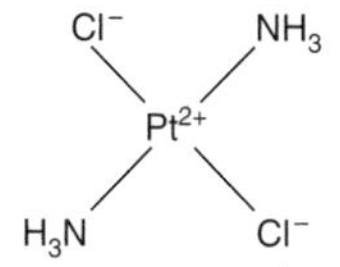

trans-platin is not

Fig. 24.9

The **name** of a complex always starts with the ligands.
Examples: $[Ag(NH_3)_2]^+$ diamminesilver(I); $[CuCl_4]^{2-}$ tetrachlorocopper(II); $[Cr(H_2O)_3(OH)_3]$ triaquatrihydroxochromate(III).

- Transition metals show **variable oxidation numbers** because the five inner d orbitals have **similar energies** to the outer 4s orbital. Metal **ions** form when the 4s and then 3d electrons are lost (usually up to M^{3+}), especially when accompanied by high lattice energy or hydration energy. **Covalent** bonds form in the usual way with the pairing of unpaired electrons, especially for the higher oxidation numbers (e.g. manganate(VII) MnO_4^-). **Dative covalent** (co-ordinate) bonds form when electron pairs are donated into vacant 3d and 4s orbitals.

Unit 24

TESTS

RECALL TEST

1 What is meant by a 'd-block element'?

(1)

2 Which subshell is filled first and emptied first, the 3d or the 4s?

(1)

3 Write the full electronic configuration of:

a Mn

b Fe^{3+} (2)

4 Why is it easy to oxidise Fe^{2+} to Fe^{3+} but difficult to oxidise Mn^{2+} to Mn^{3+}?

(3)

5 Explain why transition metals are coloured.

(4)

6 What three factors determine the colour of a particular coloured complex?

(3)

7 Explain what is meant by a complex.

(2)

8 What is a ligand?

(2)

9 Explain what is meant by a 'monodentate ligand'.

(1)

10 State the names of the eight ligands you should know, with their formulae.

(6)

11 On a piece of paper, draw EDTA. (2)

12 What is the co-ordinate number of the complex $[Cu(H_2O)_4(H_2NCH_2CH_2NH_2)]^{2+}$?

(1)

13 Why do transition metals have variable oxidation numbers?

(2)

(Total 30 marks)

CONCEPT TEST

1 Transition metals have characteristic properties.

a State what is meant by 'transition elements'.

(2)

b State which metals of the d-block (from Sc to Zn) are not transition metals.

(1)

c Explain why transition metal complexes are coloured.

____

____

____ (4)

d Explain why transition metals have variable oxidation numbers.

____

____ (2)

e Give one reason why some transition metals have high oxidation numbers in some of their compounds.

____ (1)

2 Complexes containing ligands are produced by transition metals and other metallic elements of the periodic table.

a Explain what is meant by a 'ligand'.

____ (2)

b Name these complexes:

i VO^{2+} ____

ii $CuCl_2^-$ ____

iii $Ni(CO)_4$ ____ (6)

c Complete the boxes to show the electronic configuration of iron and the iron ions:

	3d	4s	
Fe atom	(Ar) [ ] [ ] [ ] [ ] [ ]	[ ]	(Ar) indicates an equivalent configuration below this level
Fe^{2+}	(Ar) [ ] [ ] [ ] [ ] [ ]	[ ]	
Fe atom	(Ar) [ ] [ ] [ ] [ ] [ ]	[ ]	

(3)

d Why is it easy to oxidise Fe^{2+} to Fe^{3+}, but difficult to oxidise Mn^{2+} ions?

____

____ (4)

3 Some ligands are bidentate.

a Give an example of a bidentate ligand. Name it and give the structural formula.

____ (2)

b State how many of these bidentate ligands would fit around one chromium(III) ion.

____ (1)

c State the type of isomerism exhibited by the complex in part **b**.

____ (1)

d Name the shape of hexaaquachromium(III) ions.

____ (1)

(Total 30 marks)

Unit 25

TRANSITION METALS: REACTIONS

Aqueous cations	Colour
$Cr^{3+}(aq)$	purple
$Mn^{2+}(aq)$	colourless
$Fe^{3+}(aq)$	blue-green
$Fe^{2+}(aq)$	brown
$Co^{2+}(aq)$	pink
$Ni^{2+}(aq)$	green
$Cu^{2+}(aq)$	blue
$Zn^{2+}(aq)$	colourless

Fig. 25.1

Note the **more polarising** the cation (high charge + small size = high charge density), the more polarised the **water ligands**, the greater the concentration of H+ ions, and the **more acidic** the solution. Therefore, $Fe^{3+}(aq)$ is more acidic than $Fe^{2+}(aq)$. A solution of an acidic complex will evolve **CO_2** from **sodium carbonate.**

- You must know the **colours** of the aqueous cations V to Zn (see Fig. 25.1). You will find transition metal chemistry easier to understand if you learn to recognise the **types of reaction**: acid–base, redox, ligand substitution, precipitation, and thermal decomposition.

  Acid–base reactions are when a complex gains or loses a **proton** (H^+ ion). Losing a proton is called **deprotonation**. **Example:**

  $[Fe(H_2O)_6]^{3+}(aq) + H_2O(l) \rightleftharpoons [Fe(H_2O)_5(OH)]^{2+}(aq) + H_3O^+(aq)$

  Redox reactions are when an element changes oxidation state. **Example:**

  $Cr_2O_7^{2-}(aq) + 3Zn(s) + 14H^+(aq) \rightarrow 2Cr^{3+}(aq) + 3Zn^{2+}(aq) + 7H_2O(l)$

  The oxidation number of Cr reduces from +6 to +3. The oxidation number of Zn increases from 0 to +2.

  Ligand substitution (exchange or displacement) is when one ligand replaces another. **Example:** Aqueous Cu^{2+} ions are pale blue. Adding aqueous NH_3 forms a deep blue solution.

  $[Cu(H_2O)_6]^{2+}(aq) + 4NH_3(aq) \rightarrow [Cu(NH_3)_4(H_2O)_2]^{2+}(aq) + 4H_2O(l)$

  NB The reaction

  $[Cr(H_2O)_6]^{3+}(aq) + 3OH^-(aq) \rightarrow [Cr(H_2O)_3(OH)_3](s) + 3H_2O(l)$

  appears to be ligand replacement, but in fact it is an acid–base reaction in which $OH^-(aq)$ is protonated by water ligands in the complex.

  Thermal decomposition occurs as with group 2 compounds.

- Forming insoluble hydroxide **precipitates** helps to identify many transition metals. The precipitating agent is either (i) the strong base **NaOH(aq)** which contains a high concentration of $OH^-(aq)$ ions, or (ii) **aqueous NH_3** which contains small concentrations of $OH^-(aq)$ and high concentrations of $NH_3(aq)$.

  $NaOH(s) + H_2O(l) \rightarrow Na^+(aq) + OH^-(aq)$

  $NH_3(aq) + H_2O(l) \rightleftharpoons NH_4^+(aq) + OH^-(aq)$

  Generally, adding a **few drops** of either NaOH(aq) or $NH_3(aq)$ to an aqueous transition metal cation forms a **hydroxide precipitate**.

  $M^{2+}(aq) + 2OH^-(aq) \rightarrow M(OH)_2(s)$

  $M^{3+}(aq) + 3OH^-(aq) \rightarrow M(OH)_3(s)$

  You must know the colours of the aqueous ions and of the precipitates. Some **dissolve in excess** NaOH(aq) to form **anions**; some dissolve in excess $NH_3(aq)$ to form **ammine complexes**.

Cation	+ drops NaOH(aq)	+ drops $NH_3(aq)$	In excess NaOH(aq)	In excess $NH_3(aq)$
Cr^{3+}	green ppt	green ppt	green solution	does not dissolve
Mn^{2+}	grey* ppt	grey* ppt	does not dissolve	does not dissolve
Fe^{2+}	green* ppt	green* ppt	does not dissolve	does not dissolve
Fe^{3+}	brown ppt	brown ppt	does not dissolve	does not dissolve
Ni^{2+}	green ppt	green ppt	does not dissolve	violet solution of $[Ni(NH_3)_6]^{2+}$
Cu^{2+}	blue ppt	blue ppt	does not dissolve	deep blue solution of $[Cu(NH_3)_4(H_2O)_2]^{2+}$
Zn^{2+}	white ppt	white ppt	colourless solution of $[Zn(OH)_4]^{2-}$	colourless solution of $[Zn(NH_3)_4]^{2+}$

ppt = precipitate

These green* or grey* precipitates turn brown due to **oxidation** by the **air**.

$4Mn(OH)_2(s) + O_2(g) + 2H_2O(l) \rightarrow 4Mn(OH)_3(s)$

$4Fe(OH)_2(s) + O_2(g) + 2H_2O(l) \rightarrow 4Fe(OH)_3(s)$

- When chromium ions form **chromium hydroxide** and then **dissolve** in excess, the full equations are:

  (i) deprotonation
  $[Cr(H_2O)_6]^{3+}(aq) + 3OH^-(aq) \rightarrow [Cr(OH)_3(H_2O)_3](s) + 3H_2O(l)$

(ii) further deprotonation

$[Cr(OH)_3(H_2O)_3](s) + 3OH^-(aq) \rightarrow [Cr(OH)_6]^{3-}(aq) + 3H_2O(l)$

Chromium hydroxide will also dissolve in **acid**:

$[Cr(OH)_3(H_2O)_3](s) + 3H^+(aq) \rightarrow [Cr(H_2O)_6]^{3+}(aq)$

When a hydroxide **dissolves** in excess **NH_3(aq)**, an ammonia complex is formed by **ligand substitution**. **Example:**

$[Cu(OH)_2(H_2O)_4](s) + 4NH_3(aq) \rightarrow [Cu(NH_3)_4(H_2O)_2]^{2+}(aq) + 4H_2O(l)$

pale blue   dark blue

- Ligand substitution occurs when concentrated HCl(aq) (or NaCl) are added to aqueous complexes.

  $[Cu(H_2O)_6]^{2+}(aq) + 4Cl^-(aq) \rightarrow 6H_2O(l) + [CuCl_4]^-(aq)$

  Aqueous **thiocyanate ions** will form a blood-red solution with Fe^{3+}(aq) by ligand substitution (it really does look like real blood).

  $[Fe(H_2O)_6]^{3+}(aq) + SCN^-(aq) \rightarrow [Fe(SCN)(H_2O)_5]^{2+}(aq) + H_2O(l)$

  Ligands that substitute for water in **Ag^+(aq)** include: **thiosulphate** $S_2O_3^{2-}$ (dissolves unexposed AgI emulsion from photographic film to form $[Ag(S_2O_3^{2-})_2]^{3-}$(aq) during 'fixing').

- **Vanadium** chemistry usually involves **redox** reactions. (See Fig. 25.2.)

  Vanadium (**+5**) is **reduced by Zn** with 40% HCl(aq).

  If cold, to V (**+4**):

  $2VO_2^+(aq) + Zn(s) + 4H^+(aq) \rightarrow 2VO^{2+}(aq) + Zn^{2+}(aq) + 2H_2O(l)$

  Going further to V (**+3**):

  $2VO^{2+}(aq) + Zn(s) + 4H^+(aq) \rightarrow 2V^{3+}(aq) + Zn^{2+}(aq) + 2H_2O(l)$

  If boiled, to V (**+2**):

  $2V^{3+}(aq) + Zn(s) \rightarrow 2V^{2+}(aq) + Zn^{2+}(aq)$

  Note that the reaction mixture changes from (+5) yellow to green (mix of +5 and +4) to blue (+3), to violet (+2).

- There are two forms of **chromium(VI)**, depending on the pH: yellow CrO_4^{2-}(aq) and orange $Cr_2O_7^{2-}$(aq).

  $2CrO_4^{2-}(aq) + 2H^+(aq) \rightleftharpoons Cr_2O_7^{2-}(aq) + H_2O(l)$

  Cr(VI) is a **powerful oxidising agent** and is reduced by e.g. Fe^{2+}(aq), ethanol, or Zn.

  $Cr_2O_7^{2-}(aq) + 6Fe^{2+}(aq) + 14H^+(aq) \rightarrow 2Cr^{3+}(aq) + 6Fe^{3+}(aq) + 7H_2O(l)$

  Cr(III) can be **oxidised** to Cr(VI) under alkaline conditions by **hydrogen peroxide** H_2O_2 to form water and the chromate ion CrO_4^{2-}(aq).

- **Manganate(VII)** MnO_4^- is a **powerful oxidising agent**. It is reduced to Mn(II) by e.g. Fe^{2+}(aq), ethanol, or Zn.

  $MnO_4^-(aq) + 5Fe^{2+}(aq) + 8H^+(aq) \rightarrow Mn^{2+}(aq) + 5Fe^{3+}(aq) + 4H_2O(l)$

  Manganese(IV) oxide oxidises Cl^-(aq) to Cl_2 (e.g. the lab preparation of Cl_2 from HCl).

  $MnO_2(s) + 4Cl^-(aq) + 4H^+(aq) \rightarrow MnCl_2(aq) + Cl_2(g) + 2H_2O(aq)$

- **Iron(III)** is readily **reduced** to iron(II) by mild reducing agents e.g.

  $Fe^{3+}(aq) + e^- \rightarrow Fe^{2+}(aq)$

  $2Fe^{3+}(aq) + 2I^-(aq) \rightarrow 2Fe^{2+}(aq) + I_2(aq)$

  Iron(II) readily **oxidises** to iron(III). $Fe^{2+}(aq) \rightarrow Fe^{3+}(aq) + e^-$

- Aqueous **copper(II)** ions can be reduced to **copper(I)**. However, aqueous Cu(I) ions **disproportionate**.

  $2Cu^+(aq) \rightleftharpoons Cu(s) + Cu^{2+}(aq)$

  CuI forms when I^-(aq) ions reduce Cu^{2+}(aq) ions.

  $2Cu^{2+}(aq) + 4I^-(aq) \rightarrow 2CuI(s) + I_2(aq)$

As with most of the corresponding s- and p-block metal compounds, transition metal **nitrates**, **sulphates**, and **chlorides** are **soluble** in water. The insoluble silver(I) and copper(I) halides are the only important exceptions. **Carbonates** and **hydroxides** are **insoluble**.

Adding NH_3(aq) to Co^{2+}(aq) forms $[Co(NH_3)_6]^{2+}$(aq) which is **oxidised by air** to form $[Co(NH_3)_6]^{3+}$(aq).

Note the co-ordination number **change** from 6 to 4. Only four of the large Cl^- ions can fit around the small Cu^{2+} ion. You should **know** the examples $[CuCl_4]^{2-}$(aq) (yellow) and $[CoCl_4]^{2-}$(aq) (blue).

Ammonia NH_3(aq) forms $[Ag(NH_3)_2]^+$(aq) which is the active ingredient in Tollen's reagent.

Cyanide ion CN^-(aq) forms $[Ag(CN)_2]^-$(aq) used in silver plating.

Vanadium ion colours

(+5) VO_2^+(aq) – yellow
(+4) VO^{2+}(aq) – blue
(+3) V^{3+}(aq) – green
(+2) V^{2+}(aq) – violet.

You can remember this using the mnemonic 'Young Badgers Grunt Violently' (or make up your own!).

Fig. 25.2

TiO_2 is used in white paint.

Look at Fehling's reaction in unit 20 to see copper(I) reduced.

Unit 25

TESTS

RECALL TEST

1 State the colours of these aqueous ions:

Cr^{3+} ______ Mn^{2+} ______ Fe^{2+} ______
acidified Fe^{3+} ______ Co^{2+} ______
Ni^{2+} ______ Cu^{2+} ______ Zn^{2+} ______ (8)

2 What types of reaction are these?

a $5MnO_4(aq) + 8H^+(aq) + 5Fe^{2+}(aq) \rightarrow Mn^{2+}(aq) + 5Fe^{3+}(aq) + 4H_2O(l)$

______

b $[Zn(H_2O)]^{2+}(aq) + 2OH^-(aq) \rightarrow [Zn(OH)_2(H_2O)_4](s) + 2H_2O(l)$

______

c $[Zn(OH)_2(H_2O)_4](s) + 2OH^-(aq) \rightarrow [Zn(OH)_4]^{2-}(aq) + 2H_2O(l)$

______

d $[Cu(H_2O)_6]^{2+}(aq) + 4Cl^-(aq) \rightarrow 6H_2O(l) + [CuCl_4]^-(aq)$

______ (4)

3 Indicate which of these aqueous ions react as described:

	Cr^{3+}	Mn^{2+}	Fe^{2+}	Fe^{3+}	Co^{2+}	Ni^{2+}	Cu^{2+}	Zn^{2+}
Forms ppt with NaOH(aq)								
Dissolves in excess NaOH(aq)								
Forms ppt with $NH_3(aq)$								
Dissolves in excess $NH_3(aq)$								
ppt oxidises in air								

(8)

4 Write balanced equations for these reactions:

a aerial oxidation of iron(II) hydroxide

______

b chromium(III) hydroxide dissolving in excess NaOH(aq)

______

c chromium(III) hydroxide dissolving in excess HCl(aq)

______

d oxidation of Cr(III) ions by H_2O_2

______ (4)

5 Explain how the vanadium oxidation states may be decreased stepwise to vanadium(II) ions.

______

______ (5)

6 State the reagent which would oxidise vanadium(II) ions to vanadium(V) ions.

______ (1)

(Total 30 marks)

CONCEPT TEST

1 Iron is a typical transition metal.

a Iron(II) sulphate dissolves in water to make a pale green solution. On addition of NaOH(aq) a green precipitate forms, which changes colour when left in the air.

i Give the formula of the first precipitate.

_______________________________________ (1)

ii What colour would the green precipitate turn in air?

_______________________________________ (1)

iii Write a balanced ionic equation for this redox reaction.

_______________________________________ (2)

b Iron(II) ions will react with manganate(VII) ions (permanganate) to produce manganese(II) ions.

i Write a half equation for the oxidation of iron(II) ions.

_______________________________________ (2)

ii Write a half equation for the reduction reaction.

_______________________________________ (2)

iii Hence write an ionic equation for the oxidation of iron(II) ions by manganate(VII) ions.

_______________________________________ (2)

2 Pale blue aqueous copper(II) ions, $[Cu(H_2O)_6]^{2+}$, with drops of dilute aqueous NH_3 will form a blue precipitate, $Cu(OH)_2$, which redissolves in excess NH_3(aq) to form a dark blue solution, $[Cu(NH_3)_4(H_2O)_2]^{2+}$.

a State the type of reaction that is illustrated by the formation of a blue precipitate.

_______________________________________ (1)

b State the type of reaction for the formation of a dark blue solution.

_______________________________________ (1)

c How could the dark blue solution be converted back to form a pale blue solution?

_______________________________________ (2)

d How could the aqueous copper(II) ions be converted into yellow tetrachlorocopper(II) ions (tetrachlorocuprate(II) ions)?

_______________________________________ (2)

3 Which reagents will:

a reduce VO_2^+ to V^{2+}?

____

b oxidise V^{2+} to VO_2^+ or VO^{3-} ions?

____

c change CrO_4^{2-} into $Cr_2O_7^{2-}$?

____

d convert $Cr(OH)_3$(s) into $Cr(OH)_6^{3-}$ solution?

_______________________________________ (4)

(Total 20 marks)

REDOX EQUILIBRIA: $E^{\ominus}$

You need to know about **oxidation numbers** (see unit 4) to understand this section.

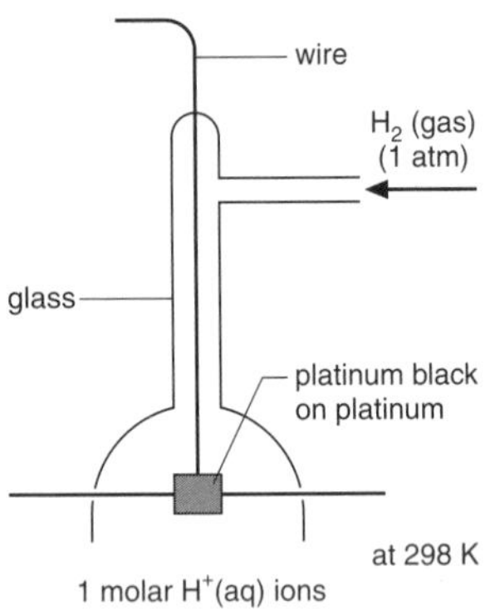

Fig. 26.1 The hydrogen electrode

- Predicting whether a reaction will happen is a powerful idea in chemistry. Whether a reaction is **feasible** can be decided from **electrode potential** $E^{\ominus}$ values (if a **redox** reaction) or from **Gibbs energy** ΔG values for **any** reaction.
- When a piece of metal (called an **electrode**) is dipped into water, some metal atoms ionise. They leave **electrons** behind in the **metal** as they enter the water as **hydrated ions. Example:** Zinc: $Zn(s) \rightarrow Zn^{2+}(aq) + 2e^-$

  Some hydrated metal ions will find their way back to the electrode and **recombine** with the free electrons on it.

  $Zn^{2+}(aq) + 2e^- \rightarrow Zn(s)$

  In time the two reaction rates will equalise and **equilibrium** will be attained.

  $Zn(s) \rightleftharpoons Zn^{2+}(aq) + 2e^-$

  A **metal** electrode in contact with a solution of its **ions** is a **half cell**. There are electrons on the electrode and positive ions in solution, so the metal will have a negative charge and the solution a positive charge. There will be a **potential difference** (p.d.) between the solution and the metal electrode. This voltage is called the **electrode potential**.

Arranging $E^{\ominus}$ values in order produces the **electrochemical series**, which shows the relative oxidising or reducing powers of substances.

- You cannot **measure** the electrode potential directly by connecting a voltmeter to a half cell. A voltmeter connecting **wire** that is dipped into the solution will have its **own electrode potential**. Therefore, the electrode potential of a metal must be compared with another electrode used as a **standard**. This arrangement is rather like measuring altitude from sea level rather than from the (unreachable) centre of the Earth. The standard **hydrogen electrode** is chosen as the standard and is assigned a potential of zero volts (see Fig. 26.1).

Powerful **oxidising agents** generally have $E^{\ominus}$ values greater than +1 V. **Example:** Fluorine, the strongest common oxidant.

$F_2(g) + 2e^- \rightleftharpoons 2F^-(aq)$

$E^{\ominus} = +2.87$ V

- The **standard electrode potential** of a half cell is the potential difference (measured in volts) between the half cell and a standard hydrogen electrode under standard conditions and when **no current flows**. Standard conditions are 298 K, 100 kPa, and solution concentration 1.00 mol dm^{-3}. The half equations for the zinc and the hydrogen electrodes are:

  $2H^+(aq) + 2e^- \rightleftharpoons H_2(g)$ 0.00 volts

  $Zn^{2+}(aq) + 2e^- \rightleftharpoons Zn(s)$ −0.76 volts compared with the H electrode.

  When a Zn half cell is connected to a hydrogen electrode and the circuit is completed, Zn produces electrons that flow to the hydrogen electrode. The potential has a **negative sign**.

  The **salt bridge** is a filter paper strip soaked in saturated aqueous KNO_3. As electrons flow through the external circuit, hydrated ions flow through the salt bridge to **complete** the circuit.

Powerful **reducing agents** generally have $E^{\ominus}$ values more negative than −1 V. **Example:** Potassium, the strongest common reductant.

$K^+(aq) + e^- \rightleftharpoons K(s)$

$E^{\ominus} = -2.92$ V

- You can **predict** whether a redox reaction is feasible by comparing electrode potentials.

  Example: Will Zn metal reduce Cu^{2+}(aq) ions? From a data book, the standard electrode potentials for the two redox reactions concerned are:

  $Cu^{2+}(aq) + 2e^- \rightarrow Cu(s)$ $E^{\ominus} = +0.34$ V

  $Zn^{2+}(aq) + 2e^- \rightarrow Zn(s)$ $E^{\ominus} = -0.76$ V

  The zinc half cell has the **more negative** potential, so electrons will flow **from the zinc** electrode **to the copper** electrode (which has a less negative i.e. more positive potential). The half equation for the zinc half cell moves to the **left** as it produces electrons; the half equation for the copper half cell moves to the **right** as it consumes electrons (see Fig. 26.2).

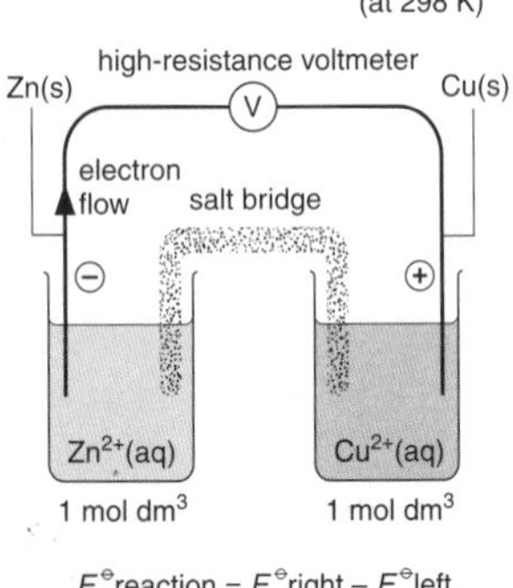

$E^{\ominus}$reaction = $E^{\ominus}$right – $E^{\ominus}$left
= +0.34 – $^-$0.76
= +1.1 volts

Fig. 26.2

The more negative $E^{\ominus}$ is the negative electrode

The more negative $E^{\ominus}$ produces e^-

⊖ $Zn^{2+}(aq) + 2e^- \rightleftharpoons Zn(s)$ −0.76 V
reactant

⊕ $Cu^{2+}(aq) + 2e^- \rightleftharpoons Cu(s)$ +0.34 V
reactant

So this is feasible: $Zn(s) + Cu^{2+}(aq) \rightarrow Zn^{2+}(aq) + Cu(s)$ and the opposite is not feasible.

$Zn(s) \rightarrow Zn^{2+}(aq) + 2e^-$ (**oxidation**) $Cu^{2+}(aq) + 2e^- \rightarrow Cu(s)$ (**reduction**)

Adding the two half equations gives the full **redox** equation:

$Zn(s) + Cu^{2+}(aq) \rightarrow Zn^{2+}(aq) + Cu(s)$

The electrode potential of an **ion pair** is measured by placing a solution of concentration 1.0 mol dm^{-3} with respect to each ion in contact with a platinum electrode. **Example:** $Fe^{3+}(aq) + e^- \rightleftharpoons Fe^{2+}(aq)$ $E^\ominus = +0.77$ V

The reaction happens either when the reactants are **separated** into two half cells or when they are **mixed** together in a test tube.

- Sometimes the electrode potentials predict that a reaction is **feasible**, but it does **not** occur. You should conclude that the reaction must have a high activation energy, so the rate is **low**. It is **kinetically stable**.

  Sometimes the standard electrode potentials predict that a reaction is not possible, but the reaction does occur in a test-tube. You must state that **non-standard conditions** are being used. Often the reactant concentrations are greater than 1.0 mol dm^{-3}.

You can apply **Le Chatelier's principle** to estimate the effect of non-standard conditions.

- Electrochemical cells may be written in a short-hand form called **cell notation** (see Fig. 26.3). Note that the species with the **higher** (**more positive**) oxidation number is written next to the salt bridge (¦). Phase boundaries are represented by |.

Pt(s) | $H_2(g)$ | $H^+(aq)$ ¦	Hydrogen electrode half cell
Pt(s) | $Fe^{3+}(aq)$, $Fe^{2+}(aq)$ ¦	Fe(II)/Fe(III) half cell
Zn(s) | $Zn^{2+}(aq)$ ¦ $Cu^{2+}(aq)$ | Cu(s)	Zinc/copper cell

Fig. 26.3

When using cell notation to represent a cell, the **hydrogen half cell** (if present) must always be written on the **left**. Otherwise, the electrode with the **more negative potential** is written on the left. As a result, the e.m.f. (voltage) of a cell that represents the **overall** reaction is given by:

$E_{reaction} = E_{right} - E_{left}$

Disproportionation may be explained by using $E^\ominus$ values.

- **Rusting** (the corrosion of iron) is an electrochemical process. In the presence of water, oxygen in the air oxidises iron to iron(II), which then forms iron(III) hydroxide, and then hydrated iron(III) oxide (rust).

		$E^\ominus$
⊖	$Fe^{2+}(aq) + 2e^- \rightleftharpoons Fe(s)$ (←)	−0.44
⊕	$\frac{1}{2}O_2(g) + H_2O(l) + 2e^- \rightleftharpoons 2OH^-(aq)$ (→)	+0.40

So $\frac{1}{2}O_2(g) + H_2O(l) + Fe(s) \rightarrow 2OH^-(aq) + Fe^{2+}(aq)$

then $Fe^{2+}(aq) + 2OH^-(aq) \rightarrow Fe(OH)_2(s)$

then $2Fe(OH)_2(s) + \frac{1}{2}O_2(g) + H_2O(l) \rightarrow Fe(OH)_3(s)$

An **electrochemical cell** consists of two connected half cells. The voltage of the cell is equal to the arithmetic difference between the $E^\ominus$ values.

- The **dry cells** you use in your personal stereo are electrochemical cells. A group of cells joined together is called a **battery**. These cells produce electricity by chemical redox reactions at their electrodes. **Oxidation** (loss of e^-) happens at one electrode while **reduction** (gain of e^-) happens at the other.

Storage cells are recharged when an externally applied current reverses the chemical changes.

- **Fuel cells** work by using oxygen (usually from the air) as the oxidising agent and a fuel (hydrogen or a hydrocarbon, e.g. methane) as the reducing agent. Most fuel cell electrodes consist of a metal foam that has a large surface area. Fuel cells are much more **energy efficient** than conventional power stations, and are portable and rugged.

Unit 26

TESTS

RECALL TEST

1 Define 'standard electrode potential'.

(3)

2 On a sheet of paper, draw and label the hydrogen electrode. (7)

3 For these two electrode potentials, work out a balanced equation for a feasible reaction, and find which electrode is positive, and what the e.m.f. might be:

$Ag^+(aq) + e^- \rightarrow Ag(s)$ +0.08 volts

$Ni^{2+}(aq) + 2e^- \rightarrow Ni(s)$ –0.25 volts

(3)

4 Write out the cell notation for the following:

a an $Ag(s)/Ag^+(aq)$ half cell connected to a hydrogen half cell

b an $Ni(s)/Ni^{2+}(aq)$ half cell connected to a $Fe^{3+}(aq)/Fe^{2+}(aq)$ half cell

(4)

5 When the electrode potentials indicate a reaction is feasible but it does not occur, what must this be due to?

(2)

6 Why may a reaction occur in the test-tube even though the standard electrode potentials suggest it should not be feasible?

(2)

7 For rusting ______ and ______ are required. (2)

8 Fuel cells use ______ or ______ and ______ as the fuel to produce electricity. (3)

9 Give three advantages of fuel cells over conventional power stations.

(3)

10 A list of the metal electrode potentials in order is called the

(1)

(Total 30 marks)

CONCEPT TEST

1 a Use the electrode potentials opposite to predict which halide ions may be oxidised by oxygen to produce the halogens.

(2)

b In fact, if oxygen is mixed with the aqueous halide ions, no halogen is immediately produced. Suggest why this might be so.

(2)

		$E^{\ominus}$
A	$Fe^{2+}(aq) + 2e^- \rightarrow Fe(s)$	–0.44
B	$Fe^{3+}(aq) + e^- \rightarrow Fe^{2+}(aq)$	+0.77
C	$Zn^{2+}(aq) + 2e^- \rightarrow Fe(s)$	–0.76
D	$Ni^{2+}(aq) + 2e^- \rightarrow Ni(s)$	–0.25
E	$I_2(aq) + 2e^- \rightarrow 2I^-(aq)$	+0.54
F	$Br_2(aq) + 2e^- \rightarrow 2Br^-(aq)$	+1.07
G	$Cl_2 + 2e^- \rightarrow 2Cl^-(aq)$	+1.36
H	$O_2(g) + 4H^+(aq) + 4e^- \rightarrow 2H_2O(l)$	+1.23

2 Iron(II) ions are easily changed into iron(III) ions by many oxidising agents. Also, some reducing agents will reduce iron(III) ions to iron(II) ions.

a Will iron(III) ions oxidise iodide ions? Explain your answer.

(2)

b In practice, aqueous iron(II) ions will reduce iodine to produce iodide ions. Explain why this occurs.

(2)

c Which reagents, if any, in the equations above will oxidise iron(II) ions to iron(III) ions? Explain your answer.

(2)

d Use the electrode potentials above to explain which oxidation number is produced by the action of oxygen with water on iron.

(2)

e Give the cell notation for the two half cells involved in rusting.

(2)

f Explain why zinc plating protects iron from rusting.

(2)

g Explain why nickel plating protects iron until the nickel is breached.

(2)

h Explain why iron(III) ions do not disproportionate.

(2)

(Total 20 marks)

Unit 27 LINKING TOGETHER CHEMISTRY

The examiners are only allowed to ask questions on topics that are on the syllabus.

- If you are taking **A2** level papers, then you will have to face the general, or **synoptic papers** that draw together the whole syllabus under the heading **Unifying Concepts**. These titles may sound frightening and it may seem that the examiner can change topics within a question without warning. However, if you fully understand the basics of chemistry and stay calm, you will find these questions an interesting challenge.

  You will face questions that may seem to have nothing to do with the syllabus you studied. If you cannot immediately think of the relevant chemistry, **don't panic**. Pause and let your mind wander the syllabus. Does the question remind you of an idea you have studied?

It is crucial that you understand the **command words** that the examiner uses. Look at the introduction for explanations of these words.

- Your examiners have agreed the following aims for **general or synoptic questions**. They state:

  "... Candidates should be able to:

  (i) bring together knowledge, principles and concepts from different areas of chemistry, including experiment and investigation, and apply them in a particular context, expressing ideas clearly and logically and using appropriate specialist vocabulary;
  (ii) use chemical skills in contexts which bring together different areas of the subject. ..."

- So be ready to:

  Use facts, ideas, and practical knowledge, from **any** part of the syllabus.
  Apply these in **new** situations.
  Suggest new ideas based on known ideas.
  Communicate your answer clearly, using **chemical** terms.

- The examiner can jump from one topic to another in any part of the syllabus. Here are some **common connections**:

  Rates connected to **mechanisms** e.g. S_N1 and S_N2.
  Thermochemistry with any organic or inorganic reactions.
  Rates, **equilibrium**, and **thermochemistry** together applied to an **industrial process** that you have not met before.
  Bond enthalpies applied to **organic chemistry** (perhaps the stages of a mechanism) and so to **reaction rates**.
  Electrode potentials with inorganic reactions, particularly those involving **transition metals** or group 7 elements.
  An **organic mechanism** applied to a new situation.
  Mechanism applied to an **inorganic** setting e.g. a covalent chloride.
  Organic chemicals used as **ligands** with d-block metal ions.
  Mole calculations mixed in with any of the above.

- A **worked example**: This question deals with various aspects of copper chemistry.

  (a) Copper ions, in a complex, are used in a test for a functional organic group.

  (i) Name the test. (1 mark)
  (ii) Name the functional group that gives a positive result in the test. (1)
  (iii) Give the physical state and formula of the copper-containing species produced. (2)

  (b) Here are some standard electrode potentials ($E^{\ominus}$) in volts (V).

A	$Cu^+(aq) + e^- \rightleftharpoons Cu(s)$	$E^{\ominus} = +0.52$ V	D	$I_2(aq) + 2e^- \rightleftharpoons 2I^-(aq)$	$E^{\ominus} = +0.54$ V
B	$Cu^{2+}(aq) + e^- \rightleftharpoons Cu^+(aq)$	$E^{\ominus} = +0.15$ V	E	$Cu^{2+}(aq) + I^-(aq) + e^- \rightleftharpoons CuI(s)$	$E^{\ominus} = +0.87$ V
C	$Cu^{2+}(aq) + 2e^- \rightleftharpoons Cu(s)$	$E^{\ominus} = +0.34$ V	F	$S_2O_8^{2-}(aq) + 2e^- \rightleftharpoons 2SO_4^{2-}(aq)$	$E^{\ominus} = +2.01$ V

  (i) Study A and B above and use the electrode potentials to explain the disproportionation of $Cu^+(aq)$ and to write a balanced equation for the reaction. (2)

(ii) Study B and F and write a balanced equation for the reaction that happens when the half cells are connected together. State the cell e.m.f. and identify the positive electrode. (4)

(iii) When aqueous copper(II) ions are mixed with aqueous iodide ions, this reaction occurs:

$2Cu^{2+}(aq) + 4I^{-}(aq) \rightarrow 2CuI(s) + I_2(aq)$

With reference to the electrode potentials, suggest why this reaction occurs. (3)

(iv) Explain why there is no observable change when aqueous persulphate ions $S_2O_8^{2-}(aq)$ are mixed with aqueous potassium iodide. (3)

(c) Why is it effectively impossible to change copper(II) compounds into copper(III) compounds? (2)

(total 18 marks)

- The answers discussed:

(a) (i) Here you need to think of a distinctive organic test that uses copper ions. Recalling all organic tests, you should think of the Fehling's test which uses a copper ammine complex.

(ii) The name of the functional group is aldehyde.

(iii) The formula of the copper compound made is Cu_2O, copper(I) oxide, which is a red solid.

(b) (i) A and B are electrode potentials. You need to recall that electrons flow from the more negative half cell to the more positive i.e. from B to A. Equilibrium B moves to the left and A to the right. Reversing B and adding to A produces:

B	$Cu^{+}(aq)$	$\rightarrow Cu^{2+}(aq) + e^{-}$	(reversed)
A	$Cu^{+}(aq) + e^{-}$	$\rightarrow Cu(s)$	
	$2Cu^{+}(aq)$	$\rightarrow Cu^{2+}(aq) + Cu(s)$	(note the electrons cancel)

(ii) F has the more positive electrode potential and so will have the positive electrode. Electrons will flow from equilibrium B to F. Equilibrium B moves to the left and F to the right. B produces only one electron while F consumes 2. **Doubling** B, reversing it, and adding to F gives

B	$2Cu^{+}(aq)$	$\rightarrow 2Cu^{2+}(aq) + 2e^{-}$	(doubled and reversed)
F	$S_2O_8^{2-}(aq) + 2e^{-}$	$\rightarrow 2SO_4^{2-}(aq)$	

$2Cu^{+}(aq) + S_2O_8^{2-}(aq) \rightarrow 2Cu^{2+}(aq) + 2SO_4^{2-}(aq)$ (electrons cancel)

(iii) One answer would be that the $E^{\ominus}$ value for D is more negative than the value for E. Equilibrium D will produce electrons (I^{-} is oxidised to I), and E will accept electrons (Cu^{2+} is reduced to Cu^{+}). Therefore, the reaction as written is feasible.

(iv) When you check the $E^{\ominus}$ values for the relevant equilibria (D and F), you conclude that the reaction is feasible. The question states that it does not occur, so you must look for another explanation. In this case, suggest that the activation energy may be high so the rate is so low that the reaction does not appear to happen.

(c) Many other metals have a maximum oxidation state of +2, so why do they not oxidise further? Think about what would be involved in achieving the change:

$Cu^{2+}(aq) \rightarrow Cu^{3+}(aq) + e^{-}$

The answer is that the third ionisation energy is very high (due to breaking into a closed electron shell) and requires the input of an enormous amount of energy unlikely to be recovered through hydration of the 3+ ion. The overall reaction would therefore be extremely endothermic.

> The examiner used the word **suggest** to check whether you really knew that the electrode potentials can explain why the reaction happens. It is always worth seeing whether an immediate and obvious explanation works first. Then, if it does not work, you must think further.

> The examiner has changed the subject (note the letter change from part (b) to (c), hinting at a new topic).

Unit 28 EXPERIMENTAL SKILLS

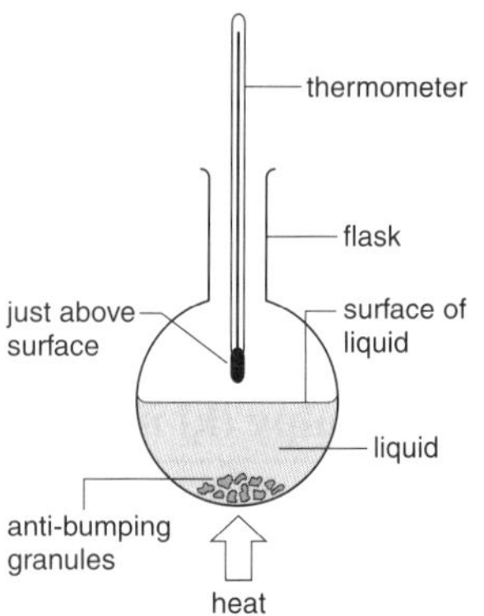

Fig. 28.1

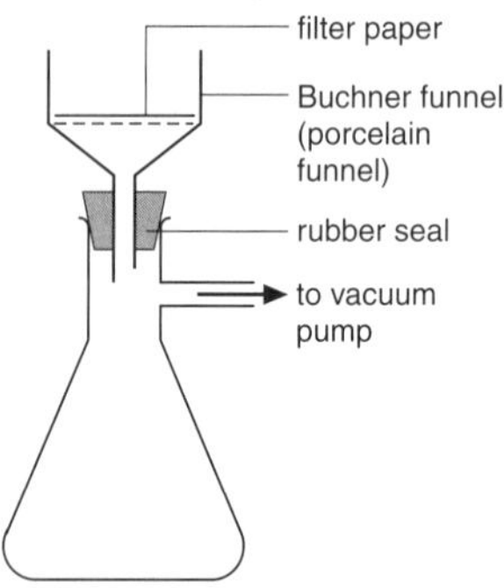

Fig. 28.2

> The two key terms associated with **filtration** are **filtrate** (the liquid that passes through the filter paper) and **residue** (the solid trapped by the filter paper).

- There is not enough space here to give detailed descriptions of the methods used in practical chemistry. The intention of these notes is to show you how to use or describe **practical techniques** in written exams or assessed practicals.

  To ensure all reactant molecules have **full contact** with each other, mixtures of liquids must be **mixed** by stirring or shaking.

  The term **heat strongly** is associated with the thermal decomposition of compounds. You should **heat gently** when trying to speed up a reaction without boiling away solvent water or causing decomposition. Use an **oil bath** to heat above 100 °C. A bare flame causes localised hot-spots. An **electric heater** is especially useful when heating flammable liquids.

  The **melting point** of a compound (up to about 250 °C with an oil bath) is determined by using a capillary tube and thermometer (see Fig. 20.7).

  The **boiling point** is measured by placing a thermometer **just above** the surface of the boiling liquid. A flask is a convenient container (see Fig. 28.1).

  Heating or boiling **under reflux** speeds up a reaction without losing volatile (readily vaporised) reactants or products. A flask with a vertical condenser attached is used (see Fig. 10.3).

  You **filter under reduced pressure** to quickly separate a solid from a liquid or to filter a hot mixture before it cools. The apparatus is a **Buchner funnel** and (side-arm) **filter flask** attached to a vacuum pump (driven by the flow from a tap) (see Fig. 28.2).

- Use **recrystallisation** to purify a compound which is more soluble in hot solvent than cold. Examiners often ask you to state the stages in recrystallisation and sometimes ask for explanations. The stages are:

  (i) **Dissolve** the impure substance (which is effectively a mixture) in the **minimum** amount of **hot solvent** (inorganic compounds in water; organic compounds in water, or ethanol, or a liquid hydrocarbon);
  (ii) **Filter** the hot mixture under reduced pressure, to remove insoluble impurities (see above);
  (iii) **Cool** the filtrate slowly to form crystals (fast cooling traps impurities within the crystals);
  (iv) **Filter** the cold mixture, to remove soluble impurities. The **residue** is the pure substance.

- Use **fractional distillation** to separate two volatile liquids. The liquid mixture is gently boiled. The vapour contains a **higher** proportion of the **more volatile** substance (the one with the **lower** boiling point). NB it is not correct to say that the more volatile substance boils first. The vapour mixture separates as it rises inside the fractionating column. The **less volatile** substance condenses and falls back as a **liquid** into the flask. The **more volatile** substance reaches the **top** of the column as a **vapour** (see Fig. 28.3).

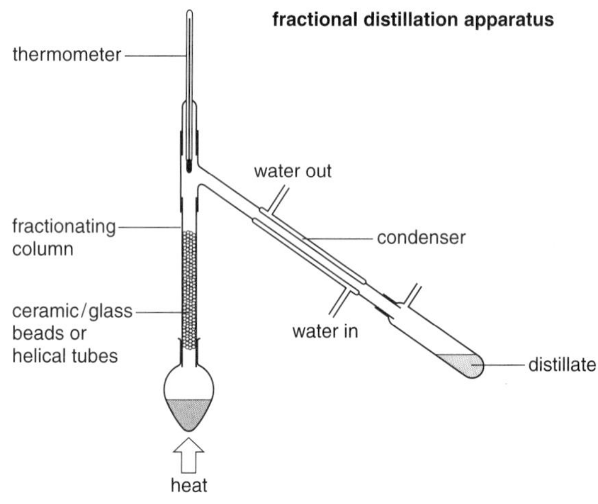

Fig. 28.3

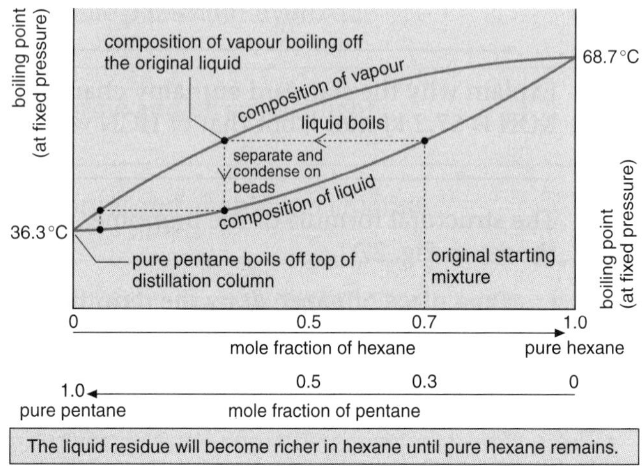

Fig. 28.4

You should be able to use a **b.p./composition** graph to explain fractional distillation (see Fig. 28.4).

- The **enthalpy change** for a reaction taking place in solution can be calculated from **temperature changes** and measurements of **mass/volume** (grams/cm^3 of water) and **amount** (moles of reactants). You must remember to state clearly the temperatures you read **before the start** of the reaction and **at the end**. Remember also that **inaccuracies** are introduced by **heat loss** to the container or to the air (by conduction or evaporation). **Incomplete reaction** may be caused by insufficient stirring.
- Details of **safety points** must be included in assessed practicals and they are often examined in written papers. The examiners assume that safety glasses and lab coats are worn, but **mention them anyway**. Watch out for **flammable** substances and include the phrase use **no naked flames** in your practical instructions. Where poisonous solids are used add the obvious phrase '**do not eat**' and include the use of protective gloves. Volatile **poisons** must be used in a **fume cupboard**.

  Precision is important. Many instruments can usually provide highly accurate data, but if you are asked to state a measurement to only 1 °C or to 2 significant figures **then do so**. The **least accurate** measurements (those with the least number of significant figures) **limit** the overall experimental accuracy. If data is correct to **2** significant figures, you **cannot** give the result of a calculation to **3** significant figures.
- You must know the **tests** for inorganic **cations** and the **flame** test colours. The addition of **NaOH(aq)** to aqueous cations is very important, so check you know the reaction of OH^-(aq) with:

  $\mathbf{Mg^{2+}}$, $\mathbf{Ca^{2+}}$, $\mathbf{Sr^{2+}}$, $\mathbf{Ba^{2+}}$, $\mathbf{Al^{3+}}$, $\mathbf{Zn^{2+}}$ and **transition metal** cations.

  Also remember that heating any **ammonium** NH_4^+ salt with NaOH(aq) produces **ammonia** NH_3 gas, e.g.

  $NH_4^+(aq) + OH^-(aq) \rightarrow NH_3(g) + H_2O(l)$
- You must also know the **tests** for inorganic **anions**.

  Halide ions form precipitates with $AgNO_3$(aq) (see unit 4 for details);
  Carbonates CO_3^{2-} produce CO_2 when added to acid (fizzing – effervescence) and when heated strongly (except group 1 carbonates).
  Hydrogencarbonates HCO_3^- (only group 1 stable as solids) such as $NaHCO_3$ also produce CO_2 when added to acid and when gently heated.
  Sulphites SO_3^{2-} produce SO_2 with acid, and form a precipitate with Ba^{2+}(aq) that dissolves in acid.
  Sulphates SO_4^{2-} form a heavy white precipitate with Ba^{2+}(aq) that is insoluble in acid.
  Hydrogensulphates HSO_4^- (as in $NaHSO_4$) react like sulphate ions. Their aqueous solutions are very acidic.
  Nitrates NO_3^- will oxidise pale blue-green aqueous iron(II) to brown iron(III) solution. Heating a solid group 1 nitrate produces nitrite and oxygen. All other solid nitrates decompose to metal oxide (or metal), nitrogen dioxide, and oxygen. Nitrates produce ammonia when heated with a strong reducing agent (e.g. zinc).
- You should know the tests for **gases**.

  Hydrogen burns with a squeaky pop when a lighted splint is applied.
  Oxygen relights a glowing splint.
  Chlorine is pale green-yellow and bleaches litmus.
  Nitrogen dioxide NO_2 is brown and bleaches litmus.
  Sulphur dioxide SO_2 turns aqueous potassium dichromate from orange to green. **Carbon dioxide** turns limewater $Ca(OH)_2$(aq) milky. The precipitate **re-dissolves** when excess gas is passed.

  $Ca(OH)_2(aq) + CO_2(g) \rightarrow CaCO_3(s) + H_2O(l)$

  $CaCO_3(s) + H_2O(l) + CO_2(g) \rightarrow Ca(HCO_3)_2(aq)$

Rate experiments usually involve measuring time. Remember to state clearly the **start time** and the **stop time**.

Remember also that solutions must be made up accurately.

Stir the liquids as soon as they are poured together.

Any of the **five main factors** (concentration, pressure, temperature, surface area, and catalyst) may influence the rate. Light may also have an influence.

Accuracy improves if the experiment is repeated and an **average** calculated.

Unit 28 PLANNING PRACTICALS

It is quite easy to become very confused and stressed when **planning** practical work for assessment or **describing** a practical in detail during a written exam. Do the planning in the following order to establish a **framework**:

(A) Read the **instructions** given.

(B) Outline the **chemistry** involved.

(C) Outline a **method** (without amounts or descriptions of apparatus).

(D) Read the instructions again (have you **missed** something?).

(E) Calculate suitable **amounts** in moles of the reactants.

(F) Now choose your **apparatus**.

(G) Remember to **stir** or **shake** mixtures. State **when** to make **measurements** such as time or temperature (start and finish). Are you **heating**: if so, how strongly? **Repeat** the experiment keeping all independent variables constant and calculate an **average** for each set of measurements.

(H) **Safety** points.

(I) **Summarise** the practical procedure, give it the title **Practical Outline** and put it at the beginning of your written account.

(J) State the **assumptions** you are making (about e.g. specific heat capacity, heat loss, purity of chemical reagents).

(K) **Accuracy** must usually be assessed. State all sources of error and indicate which are the main ones. Assumptions are often sources of errors.

(L) **Re-read** the instructions, including those that seem trivial or obvious.

A good place to start is 2 cm^3 of liquid reactants or 1–2 g of solid reactant. Aim for solution concentrations from 0.01 to 0.2 $mol\,dm^{-3}$.

TESTS

RECALL TEST

1 On a sheet of paper, draw a simplified graph to show how fractional distillation occurs in steps within the apparatus. (2)

2 State how ammonium ions may be detected in solution.

________________________________ (3)

3 Which substances/ions are indicated by these observations?

a It forms a cream ppt with $AgNO_3(aq)$

____

b It effervesces with HCl(aq), and limewater turns milky

____

c It forms a white ppt with $Ba^{2+}(aq)$, which dissolves in HCl(aq), and produces a gas which turns orange dichromate ion green

____

d It forms a white ppt with $Ba^{2+}(aq)$, which does not dissolve in HCl(aq)

____

e A solid heated strongly produces a brown gas and a gas that relights a glowing splint

________________________________ (5)

(Total 10 marks)

CONCEPT TEST

1 Aqueous sodium hydrogen sulphite may be used to detect a carbonyl compound in analysis. Aldehydes and ketones will form a white precipitate with $NaHSO_3(aq)$. Once the white solid formed is purified, its melting point is determined. The carbonyl compound originally present may be ascertained by reference to a data book which lists the melting points of sodium hydrogen sulphite derivatives.

a Explain how the white solid may be recrystallised.

(4)

b Explain how to determine the melting point of the purified solid.

(3)

c The boiling point would help to confirm which carbonyl compound was present. Describe how you would determine the boiling point.

(2)

2 White solid X dissolves in water to form a colourless solution. A sample of this solution when mixed with aqueous sodium sulphate forms a heavy white precipitate Y. When solid X is heated strongly, a brown gas Z evolves and a glowing splint readily relights. The white solid produces a pale green flame colour.

White solid P dissolves in water. The solution when mixed with hydrochloric acid produces a colourless gas Q that will turn aqueous acidified potassium dichromate from orange to green. When the solid P is put into a flame a pale lilac colour is produced, which is visible through blue glass.

a State the formula of:

X

Y

Z (6)

b State the formula of:

P

Q (2)

c State how you would confirm the presence of bromide ions in a solution.

(3)

(Total 20 marks)

Unit 29

CHEMICAL CALCULATIONS

One mole of a substance contains the same number of particles as one mole of any other substance.

- The **amount of substance** (symbol n) is measured in moles. **One mole** (1 mol) is the amount of a substance that contains the same number of particles as there are atoms in exactly 12 g of carbon-12.

  Examples: 1 mol of water H_2O contains 1 mol of water molecules, 2 mol of hydrogen atoms, and 1 mol of oxygen atoms.

  1 mol of magnesium chloride contains 1 mol of Mg^{2+} and 2 mol of Cl^- ions.

The mass of 1 mol of water is 18.015 28 g i.e. [(2 × 1.007 94) + 15.9994] g.

- The **mass** of one mole of atoms of an **element** (e.g. Ne, Fe, Cl, but not Cl_2) is equal to the relative atomic mass RAM in grams.

  Example: The mass of one mole of Cl **atoms** is 35.5 g.

- The **mass** of one mole of a **molecular compound** is equal to the formula mass (i.e. the sum of the RAM values) in grams.

  Example: The mass of one mole of Cl_2 molecules is 35.5 × 2 = 71.0 g.

  Ionic compounds consist of separate ions. It is not correct to assign a relative molecular mass to these substances. You should refer to the **relative formula mass M_r** (i.e. the sum of the RAM values) in grams.

  Example: The mass of 1 mol of magnesium chloride is 95.2104 g i.e. [24.3050 + (2 × 35.4527)] g.

  The **simplest** approach is to refer to all substances in terms of their **formula masses**. The formula mass is the mass of 1 mol.

  Example: The formula mass of magnesium chloride $MgCl_2$ is 95.2104 g.

When **stuck** for ideas, try converting the **given quantities** (masses, volumes/concentrations) into **amounts** in moles. Most of the mathematical expressions work via moles. Usually one calculation links to another via the **reacting ratios** shown by the balanced chemical equation.

- You may find you can work out simple cases in your head, but it is safer in the long run to remember an equation.

$$\textbf{amount (moles)} = \frac{\textbf{mass (grams)}}{\textbf{formula mass (grams)}} \quad \text{i.e. MOLES} = \frac{\text{mass in grams}}{\text{mass of one mole}}$$

If desperate,

$$\text{moles} = \frac{\text{GRAMs}}{\text{RAMs}}$$

where RAM is relative atomic mass(es).

- The **number** of atoms in 1 mol of carbon-12 (12 g exactly) is 6.023×10^{23} and is called the **Avogadro constant** (symbol L).

- You must know how to **calculate** the **number of particles** (atoms, molecules, or ions, etc.) in a given mass of a substance. One mole of a substance (equivalent to its formula mass) contains L particles.

  Example: 1 mol of H_2O (18.015 28 g) contains 6.023×10^{23} H_2O molecules, $2 \times 6.023 \times 10^{23}$ H atoms, and 6.023×10^{23} O atoms.

1 mol of Ne (20.1797 g) contains 6.023×10^{23} Ne atoms.

1 mol of $MgCl_2$ (95.2104 g) contains 6.023×10^{23} Mg^{2+} ions and $2 \times 6.023 \times 10^{23}$ Cl^- ions.

- The **molar concentration** of a solution describes the amount of solute dissolved in a given volume of solution (usually 1 dm^3). You will have to calculate the **molar concentration** of solutions or, if given a concentration, calculate the moles of solute present. The units of concentration are mol dm^{-3} i.e. $\frac{\text{mol}}{\text{dm}^3}$ so it should not be difficult to remember that

$$\textbf{concentration (mol dm}^{-3}\textbf{)} = \frac{\textbf{amount of solute (mol)}}{\textbf{solution volume (dm}^3\textbf{)}}$$

If desperate,

$$\textbf{conc.} = \frac{\textbf{mols}}{\textbf{vols} \text{ (in dm}^3)}$$

- 1 decimetre cubed (1 dm^3) = 1 litre (1 l) = 1000 millilitres (ml)

1 dm^3 = 1000 cm^3 = 1000 ml	$dm^3 = \frac{cm^3}{1000}$

Some students prefer to remember mol = conc. × vol.

- You will be expected to **convert** a **volume** of gas into an **amount** in moles, or to convert an amount of a gas in moles into its corresponding volume. You will be told that one mole of gas occupies 24 dm^3 at room temperature and pressure (or some other volume for different conditions). This volume is called the **molar gas volume**. As two moles of a gas would occupy twice the volume of one mole, then it is simple to remember that:

  volume of gas = number of moles × volume of one mole of gas

Never use the symbol M as an abbreviation for mol. It is an obsolete term which means mol dm^{-3}.

Use dm^3 and cm^3 for all calculations.

- The **ratio** of the **volumes** of gases in a reaction is equal to the ratio of the **amount** (number of moles) of each gas in the balanced equation.

  Example: When 100 cm^3 of methane burns in air, the volumes of the gases involved may be calculated from the balanced equation. Note that all volumes are measured at the same temperature and pressure:

  $CH_4(g) + 2O_2(g) \rightarrow CO_2(g) + 2H_2O(g)$
  100 cm^3 200 cm^3 100 cm^3 200 cm^3 (of steam)

At room temperature the volume of water would be negligible because of condensation.

- You will often have to calculate the **percentage yield** of a reaction.

$$\textbf{percentage yield} = \frac{\textbf{actual amount (mol)}}{\textbf{maximum possible amount (mol)}} \times \textbf{100\%}$$

  Sometimes, one of the reactants will be in **excess**. When calculating yield, be sure to first calculate the amounts (in mol) of all the reactants and base the theoretical yield on the reactant that is present in the **smallest amount**.

Usually an organic compound undergoing reaction is not in excess; in redox reactions, the acid is in excess.

- Occasionally, you will need to derive a mass or a volume from a given **density** value. The SI unit of density is kg m^{-3}, which is equivalent to the more practical unit of g cm^{-3}.

$$\textbf{density (g cm}^{-3}\textbf{)} = \frac{\textbf{mass (g)}}{\textbf{volume (cm}^3\textbf{)}} \quad \text{i.e. mass = density} \times \text{volume}$$

Many students find the **calculation map a** useful way to see how the different equations link (see Fig. 29.1).

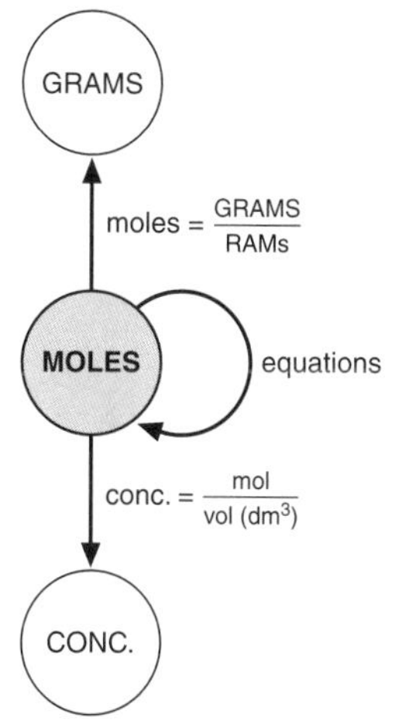

You could add other calculation steps to this diagram.

Fig. 29.1

- In titration calculations, you are given (i) the concentrations and volumes of solution X in one piece of glassware (burette or pipette), and (ii) the volume of solution Y of unknown concentration in the other piece of glassware. Work in three stages:

  (A) Calculate the moles of substance present in solution X above.

  (B) Use the balanced equation for the reaction to find the molar ratios of the reacting substances X and Y and hence the amount in moles of substance Y present in (ii).

  (C) Use the amount of Y in moles present and the volume of its solution to calculate the concentration. Of course, the examiners set this calculation regularly, so they are likely to add an extra twist, such as including a dilution.

A sample calculation:
Potassium permanganate concentration 0.020 mol dm^{-3} was used to standardise a solution of iron(II) sulphate. The titration flask held a 25.0 cm^3 **aliquot** (the solution measured by pipette) of iron(II) sulphate and 25.0 cm^3 of sulphuric acid 0.1 mol dm^{-3}. The average **titre** (volume delivered by the burette) was 27.93 cm^3. Write an ionic equation for the reaction between iron(II) sulphate and potassium permanganate. Calculate the iron(II) sulphate concentration.

A **structured** approach to **problem solving** is to **first** extract the data and **then** look for combinations of data that allow you to use an appropriate equation.

The **ionic equation** is

$MnO_4^-(aq) + 5Fe^{2+}(aq) + 8H^+(aq) \rightarrow Mn^{2+}(aq) + 5Fe^{3+}(aq) + 4H_2O(l)$

Extract the data:

$KMnO_4$ concentration = 0.02 mol dm^{-3}; vol. = 27.93 cm^3.

$FeSO_4$ concentration unknown; vol. = 25.0 cm^3.

Calculation:

(A) Amount $KMnO_4$ (mol) $= 0.02 \times \frac{27.93}{1000} = 0.000\,5586$ mol

(B) Molar ratios: 1 mol $MnO_4^-(aq)$ reacts with 5 mol $Fe^{2+}(aq)$

Amount of $FeSO_4$ (mol) = 0.002 793 mol

(C) Concentration of $FeSO_4$ (mol dm^{-3}) $= \frac{0.002\,793}{25/1000} = 0.111\,72$ mol dm^{-3}

The data is given to **two significant figures or better**, so the answer must be given to the **same** degree of accuracy i.e. concentration $FeSO_4$ = 0.11 mol dm^{-3}.

A solution of known concentration that is used to determine the concentration of another solution is a called a **standard solution**. Also, the volume and concentration of the sulphuric acid is given purely to distract you. It is in excess and is present to provide the H^+ ions required by the redox reaction.

TESTS

RECALL TEST

1 a Give an equation for moles, using mass and RAM.

(1)

b Give an equation for mass, using moles and RAM.

(1)

c Give an equation for RAM, using moles and mass.

(1)

2 a Give an equation for concentration (in mol dm^{-3}).

(1)

b Give an equation for moles using concentration and volume.

(1)

c Give an equation for volume (cm^3) using concentration and moles.

(1)

3 a If one mole of gas occupies 24 dm^3, calculate the gas volume of:

i 3 mol methane ______ **ii** 0.5 mol oxygen ______ (2)

b How many moles are there in 1000 dm^3 (1 m^3) of methane?

(1)

4 Give the equation for percentage yield.

(1)

(Total 10 marks)

CONCEPT TEST

1 During a titration of ethanoic acid against potassium hydroxide, the titre was found to be 23.45 cm^3 ethanoic acid against a 10.00 cm^3 KOH aliquot. If the alkali was 0.60 mol dm^{-3}, what is the concentration of the ethanoic acid?

$$CH_3COOH(aq) + KOH(aq) \rightarrow CH_3COO^-K^+(aq) + H_2O(l)$$

(2)

2 Often iodine is titrated against thiosulphate. Write an equation for the reaction. If a titre of 26.85 cm^3 potassium thiosulphate solution, 0.020 mol dm^{-3}, reacted with 50.0 cm^3 iodine, dissolved in potassium iodide, what would be the iodine concentration?

(3)

3 Silver chloride is less soluble than silver chromate, so the chloride ion concentration may be determined by titrating aqueous silver nitrate into a chloride ion solution. A few drops of yellow potassium chromate, here being used as an indicator, will produce a red silver chromate precipitate (mixed in with the yellow). If the aliquot was 25.0 cm^3 and the 0.010 mol dm^{-3} aqueous silver nitrate titre was 23.45 cm^3, what was the chloride ion concentration?

(3)

4 Ammonium sulphate, $(NH_4)_2SO_4$, is used as a fertiliser. A 24.0 g sample of the fertiliser was added to excess NaOH(aq) and the resulting ammonia gas absorbed in 1 dm^3 sulphuric acid, 0.10 mol dm^{-3}. 10.0 cm^3 of the resultant solution was titrated with 0.010 mol dm^{-3} NaOH. 29.10 cm^3 was required. Calculate the purity of the ammonium sulphate.

(3)

5 When 25.00 cm^3 of a copper ion solution was mixed with excess potassium iodide solution, the iodine produced required 40.10 cm^3 of 0.0200 mol dm^{-3} sodium thiosulphate solution. The reactions were these:

$$2Cu^{2+}(aq) + 4I^-(aq) \rightarrow 2CuI(s) + I_2(\text{in aqueous KI})$$

$$2S_2O_3^{2-}(aq) + I_2(aq) \rightarrow S_4O_6^{2-}(aq) + 2I^-(aq)$$

Calculate the concentration of the copper ion solution.

____ (3)

6 Benzene (M_r = 78) may be nitrated with concentrated nitric acid at 40 °C to make nitrobenzene (M_r = 123):

$$C_6H_6(l) + HNO_3(l) \rightarrow C_6H_5NO_2 + H_2O(l)$$

a A student used 10.0 grams benzene and obtained 9.0 grams nitrobenzene. What percentage yield was produced?

____ (1)

b Explain why the same student using the same amount could appear to make a yield of 120% later the same day (assuming the mathematics was correct).

____ (2)

7 During an exam practical or asessment, you were supplied with the solutions on the right:

i an aqueous solution of 3.16 g dm^{-3} potassium manganate(VII), $KMnO_4$, labelled solution P,

ii an aqueous solution of ethanedioic (oxalic) acid, $H_2C_2O_4$, of unknown concentration, labelled Q,

ii an aqueous solution of 7.865 g dm^{-3} vanadium(III) chloride, VCl_3, labelled R,

iii aqueous bench sulphuric acid.

1. Place P in a burette. Using a pipette, transfer 25.00 cm^3 of solution Q into a conical flask. Using a measuring cylinder add 25 cm^3 sulphuric acid. Warm by holding the flask over a Bunsen flame until the solution steams. Titrate the contents with solution P.

2. Using a pipette, transfer 25.00 cm^3 of solution R into a conical flask. Using a measuring cylinder add 25 cm^3 sulphuric acid and titrate the contents with solution P.

In Part 1, the average titre was 20.41 cm^3. In Part 2, the titre was 12.55 cm^3.

a Given that the following reaction takes place:

$$2MnO_4^-(aq) + 5C_2O_4^{2-}(aq) + 16H^+(aq) \rightarrow 2Mn^{2+}(aq) + 10CO_2(g) + 8H_2O(l)$$

calculate the concentration, in mol dm^{-3}, of ethanedioic (oxalic) acid in solution Q.

____

b Using the titre in Part 2, calculate how many moles of potassium manganate(VII) (permanganate) was in the titre.

____

c How many moles of vanadium(III) chloride reacted with the permanganate ions in the titre?

____

d What is the ratio between the moles of potassium manganate(VII) (permanganate) and vanadium(III) chloride?

____

e Write an equation for the reaction of vanadium(III) ions with the manganate(VII) (permanganate) ions, using the results of your calculations.

____ (8)

(Total 25 marks)

ANSWERS

UNIT 1

RECALL TEST

1 Electronic configuration. (1)
2 Electronegativity increases across the row as proton number increases. (2)
3 As atomic number increases within a group, electronegativity decreases as shells of electrons increase (so there is more shielding and a greater distance between the nucleus and the outer electrons). (3)
4 AgI is covalent because Ag and I are similar in electronegativity. (1)
5 Bonds are polar when the atoms in an covalent bond have different electronegativities. (1)
6 Van der Waals forces, permanent dipole or dipole–dipole, hydrogen bonding. (1)
7 The forces between the chlorine molecules are weak Van der Waals forces. (The covalent bonds are strong inside the molecule). (1)
8 As iodine atoms are large there are strong Van der Waals forces between the I_2 molecules. (1)
9 The Cl atom is more electronegative than the C atom. (1)
10 H bonds need an electron-deficient H (slightly positive), because the H is joined to a very electronegative atom, and NOF, which are small, very electronegative atoms. (4)
11 Ethanol has H bonds, but ethanal only has a permanent dipole. (2)
12 NaCl(l) has free ions. In NaCl(s) the ions are not free to move. (2)
13 The size of the atoms increase so the Van der Waals forces increase so the b.p. increases. (1)
14 Water H-bonds; the rest of the hydrides only have Van der Waals forces. (1)
15 The chlorine molecules lose kinetic energy, when the gas (separate randomly moving molecules) is cooled. Then the Van der Waals forces are strong enough to hold the chlorine molecules together in a liquid (touching randomly moving molecules). (4)
16 When solid NaCl (regular structure, ions touching) is heated the kinetic energy of the ions increases, until the electrostatic forces between ions is overcome, forming a liquid (ions randomly moving past each other, but touching). (4)

(Total 30 marks)

CONCEPT TEST

1 a Electronegativity is a measure of how attractive atoms are for a pair of electrons (in a covalent bond). (2)
b Carbon dioxide has covalent bonding, a sharing of electrons. (2)
c The -OH groups in glucose hydrogen-bond. Glucose dissolves in water because water can hydrogen-bond to the glucose. (2)
d i Be^{2+} is much smaller than the Ca^{2+}. (1)
ii Aluminium chloride is covalent because the Al^{3+} ion is very small and highly charged so polarises the chloride ion. (2)
iii The chloride ion is larger than the fluoride ion because the chloride ion has one more electron shell. (2)
iv The F^- ion is very small and has a single charge so cannot be polarised by the aluminium ion. (1)
2 a Graphite is made of layers of covalently bonded carbon atoms. The layers are held together by weak Van der Waals forces, so the layers can slip over each other. (2)
b All the atoms in diamond are strongly bonded together in one giant covalent lattice. (2)
c C_{60} is a solid at room temperature, because the molecules are large so the Van der Waals forces between the molecules will be strong. As the particles would slide over each other it must be slippery. (2)
d KC_{60}^+ ions with Cl^- ions must be ionic so have a high boiling point, because they are held together by strong electrostatic forces. (2)

(Total 20 marks)

UNIT 2

RECALL TEST

1 Neutron (no charge), proton (+). (If electrons are mentioned then no marks.) (2)
2 Vaporisation; ionisation; focus and acceleration; the magnetic field deflects the ions, separating them by mass and charge; the detector counts the ions of a particular mass/charge ratio; the vacuum pump removes the air to ensure ions are not deflected by air molecules. (6)
3 a Atoms of the same element (or atomic number) with different mass numbers (or numbers of neutrons). (2)
b The weighted average mass of atoms of an element (in a sample of the element) divided by 1/12th of the mass of an atom of the carbon-12 nuclide. (2)
c The number of protons and neutrons in the nucleus. (2)
4 The peak on the right. The one with the highest mass/charge ratio. (1)
5 Water: V-shaped, ammonia: pyramidal, methane: tetrahedral, beryllium chloride: linear, boron trifluoride: trigonal planar, sulphur hexafluoride: octahedral, phosphorus pentachloride: trigonal bipyramidal. (7)
6 Tetrahedral. 109.5°. (2)
7 Pyramidal. (1)
8 Methane = 109.5°, ammonia = 107°, water = 105°, carbon dioxide = 180°. (4)
9 The electron pairs (bond and lone pairs) repel until they are as far apart as possible. (1)

(Total 30 marks)

CONCEPT TEST

1 a i Using a magnetic field. (1)
ii Without a vacuum the ions would be deflected by gas molecules and not get to the detector. (1)
iii 107.972. (2)
b i $M_r = 60$. (1)
ii 15 = CH_3^+; 28 = CO^+; 45 = CO_2H^+; 60 = $C_2H_4O_2^+$ (lose 1 mark if no + charge on each ion). (4)
iii CH_3COOH, $HOCH_2CHO$. (2)
iv CH_3COOH because the mass/charge ratio of 45 is not possible for $HOCH_2CHO$. (2)
2 a 30 neutrons and 28 protons. (Maximum two marks if electrons are mentioned.) (3)
b RAM = 58.768. (2)
3 a $AlCl_3$ has 3 bonds and no lone pairs, while NH_3 has 3 bonds and one lone pair. The electron pairs repel until they are as far apart as possible. (3)
b i Draw pyramidal for PH_3 (it has 3 bonds and 1 lone pair).
ii Draw a V-shaped molecule for SO_2 (it has 2 sets of double bonds and 1 lone pair).
iii Draw pyramidal for ClO_3^- (Cl has 2 double bonds to Os, 1 single bond to an O^- ion, and has one lone pair).
iv Draw a shape based on trigonal bipyramidal for BrF_3^- (the Br is surrounded by 3 bonds to F, and two lone pairs). You could arrange the lone pairs in any direction as long as the shape looks like it is based on trigonal bipyramidal. (4)
c Octahedral. (1)
d Bond angles: CH_4 =109.5°; NH_3 = 107°; H_2O = 105°. The electron pairs repel until they are as far apart as possible. The lone pairs repel more than the bonding pairs, so with more lone pairs the bonding pairs are pushed together. (4)

(Total 30 marks)

UNIT 3

RECALL TEST

1 Magnesium atoms have one more proton than sodium atoms, and there are two electrons per atom involved in the metallic bonding compared with one electron per atom for sodium atoms. (2)
2 Mg colourless, Ca orange-red, Sr red, Ba pale or apple green, Li red, Na yellow, K lilac or pale purple. (7)
3 The potassium flame is visible through (cobalt) blue glass. (1)
4 Barium is less electronegative than magnesium. (1)
5 a $Mg(s) + 2HCl(aq) \rightarrow MgCl_2(aq) + H_2(g)$.
b $MgO(s) + 2HCl(aq) \rightarrow MgCl_2(aq) + H_2O(l)$.
c $MgCO_3(s) + 2HCl(aq) \rightarrow MgCl_2(aq) + CO_2(g) + H_2O(l)$.
d $Mg(s) + Cl_2(g) \rightarrow MgCl_2(aq)$.
e $2Mg(s) + O_2(g) \rightarrow 2MgO(s)$. (5)
6 The small double-charged Mg^{2+} ion polarises the peroxide anion, making it unstable. (2)
7 With increasing atomic number, the group 2 sulphates become less soluble. (1)
8 Because the group 2 ions become larger, so the hydroxide ion is less attractive, so the lattice energy decreases, making the hydroxides more soluble. (3)
9 Add aqueous barium chloride (or nitrate) and a heavy white precipitate will form which will not dissolve in dilute hydrochloric acid (or nitric acid), if sulphate ions are present. (2)
10 $Ca(OH)_2(aq) + CO_2(g) \rightarrow CaCO_3(s) + H_2O(l)$. (1)
11 The lithium ion is very small so polarises the nitrate anion, making it unstable. (2)
12 $BaCO_3(s) \rightarrow BaO(s) + CO_2(g)$. (1)
13 $2Na_2O_2(s) \rightarrow 2Na_2O(s) + O_2(g)$. (1)
14 $BaO_2(s) + H_2O(l) \rightarrow BaO(s) + H_2O_2(aq)$. (1)

(Total 30 marks)

CONCEPT TEST

1 a When the sodium atoms are heated the electrons are promoted to higher energy levels. When the electrons fall back to lower energy levels visible light is emitted. (3)
b Caesium reacts more vigorously than sodium, making the direct combination of Cl_2 with Cs dangerous. The Cs would react with the air. (2)
c $NaCl(s) + (aq) \rightarrow Na^+(aq) + Cl^-(aq)$. (1)
d There is a greater difference between the electronegativities of Na and I than there is between Li and I. (3)
2 a The group 2 sulphates become less soluble as the atomic number increases, because the cation radii increase so the cations become less attractive to the water molecules, so the hydration energy decreases. (The large sulphate anion means that the lattice enthalpy of the sulphates hardly differ.) (4)
b Sulphate anion, SO_4^{2-}. (1)
c As the atomic number increases the cation radii increase so the lattice energy decreases, as the larger cations keep apart the small fluoride anions. (3)
d The magnesium ions are larger than the barium ions, so the Mg^{2+} ions polarise the carbonate anions more than the Ba^{2+} ions. (2)
e The Al^{3+} ions have a greater charge and are smaller than the Mg^{2+} so polarise the CO_3^{2-} anion more, so carbonates will decompose at lower temperatures. (1)

(Total 20 marks)

UNIT 4

RECALL TEST

1 F_2 yellow, Cl_2 green, Br_2 red, I_2 black. (2)
2 Iodine molecules are large so the Van der Waals forces are strong. Chlorine molecules are small so the Van der Waals forces are weak. (2)
3 The atoms are small with few electron shells so the attraction between the nuclei and the shared electrons is strong. (2)
4 Chlorine is very reactive because of the energy released when forming ionic bonds, as the Cl electron affinity is very high, and the formation of Cl covalent bonds with other atoms is very exothermic. Iodine atoms are larger than chlorine atoms so the iodine electron affinity and covalent bonds release less energy. Iodine forms iodide ions which are large, releasing smaller lattice energies. (2)
5 a $2Fe(s) + 3Br_2(l) \rightarrow 2FeBr_3(s)$.
b $H_2(g) + Br_2(g) \rightarrow 2HBr(g)$. (2)
6 Sodium *hydrogen* sulphate, hydrogen chloride. (2)
7 Name *or* formula. I_2, $KHSO_4$, KI, S, H_2S, SO_2. (3)
8 **MgO** +2, $\mathbf{SO_2}$ +4, $H_2\mathbf{S}O_3$ +4, $\mathbf{S}O_3$ +6, $H_2\mathbf{S}O_4$ +6, $Mg\mathbf{S}O_4$ +6, $H_2\mathbf{S}$ −2, $\mathbf{N}H_3$ −3, $\mathbf{N}H_4^+$ −3, **Na**(s) 0, $\mathbf{Cl_2}$(g) 0. (11)
9 a $2Br^-(aq) + Cl_2(aq) \rightarrow Br_2(l) + 2Cl^-(aq)$.
b No reaction.
c $Ag^+(aq) + Cl^-(aq) \rightarrow AgCl(s)$.
d $PCl_5(s) + H_2O(l) \rightarrow POCl_3(aq) + 2HCl(g)$ or $PCl_5(s) + 4H_2O(l) \rightarrow H_3PO_4(aq) + 5HCl$. (4)

(Total 30 marks)

CONCEPT TEST

1 a i Aqueous brine. (1)
ii $2Cl^-(aq) \rightarrow Cl_2(g) + 2e^-$. (2)
b Disproportionation is the simultaneous oxidation and reduction of the same element. (2)
c $3Cl_2(g) + 6OH^-(aq) \rightarrow ClO_3^-(aq) + 5Cl^-(aq) + 3H_2O(l)$. Cl_2 0; ClO_3^- +5; Cl^- −1. (4)
d $ClO_3^- + 2OH^-(aq) \rightarrow ClO_4^- + H_2O(l) + 2e^-$. (2)
e Chloride ions form a white precipitate with aqueous silver nitrate, which dissolves in dilute ammonia. (2)
2 a $2S_2O_3^{2-}(aq) + I_2(aq) \rightarrow S_4O_6^{2-}(aq) + 2I^-(aq)$. (2)
b Starch makes a dark colour with iodine which otherwise would be very pale near the end point. (2)
c Add aqueous silver nitrate; a pale yellow precipitate would appear with I^-. (1)
d The green gas becomes colourless, and a orange/brown solution or a black solid forms. (2)

(Total 20 marks)

UNIT 5

RECALL TEST

1 An 'atomic orbital' is the volume around a nucleus which is occupied by an electron 95% of the time. (2).
2 **a** to **d** See text. (4)
3 Kr = $1s^2\ 2s^2\ 2p^6\ 3s^2\ 3p^6\ 3d^{10}\ 4s^2\ 4p^6$. (2)
4 An s-block element is one where the last electron added is to an s orbital. (1)
5 The electronic configuration. (1)
6 The proton number increases across the row of the periodic table. (1).
7 The proton number increases across the row. (1)
8 a Increases.
b Decreases.
c Decreases. (3)
9 The energy released when one mole of electrons is gained by one mole of gaseous atoms to form one mole of gaseous ions with a single negative charge. (1)
10 1st EA is exothermic and has a low magnitude as the nucleus in the neutral atom attracts the electron. 2nd EA is greatly endothermic as the negative ion repels the electron. (2)
11 The energy change when one mole of gaseous 3+ ions loses one mole of electrons to form one mole of gaseous 4+ ions. (2)
12 a He. (1)
b He. (1)
c Al or B. (1)
d F. (1)

13 The metals Na, Mg, and Al have high melting points due to the metallic bonding (2). Si has giant covalent structure so the highest melting point (2), and the others form small molecules which are held together by weak Van der Waals forces. (2)

(Total 30 marks)

CONCEPT TEST

1 **a** $Cl(g) \rightarrow Cl^+(g) + e^-$. (2)

b The electron lost from a He atom is from an unshielded shell (1) and the electron is held by two protons in the nucleus. (1)

c 1st EA of S: $S(g) + e^- \rightarrow S^-(g)$
2nd EA of S: $S^-(g) + e^- \rightarrow S^{2-}(g)$. (2)

d To form a metal sulphide the sulphur atom must gain two electrons, which costs energy, supplied by the release of lattice energy due to the attraction between the oppositely charged ions. (3)

2 **a** C has higher IE; C is smaller/has fewer shells than Si.

b Ar has a higher IE; the electron is lost from a shell closer to the nucleus.

c Be has higher IE; the electron lost from Be is from an s orbital, closer to the nucleus than the electron lost from the p orbital in B.

d Mg has higher IE; Mg has one more proton than Na.

e Na has higher IE; the electron is lost from a shell closer to the nucleus. (10)

3 **a** The graph should show the IE generally increases as electrons are lost, with jumps after the 1st, 9th, and 17th electrons are lost. (4)

b The electron is lost from a cation with a greater charge. (1)

c The graph reflects the electronic structure of a potassium atom: 2, 8, 8, 1. The first e^- is lost from the outer shell, then the next 8 from one shell in, which is closer to the nucleus so the IE will be larger. The next 8 e^- is from the next shell in. The last 2 e^- are in the innermost shell so require the largest amount of energy. (4)

d K has only 19 electrons. (2)

(Total 30 marks)

UNIT 6

RECALL TEST

1 Elements: Na, Mg, Al, Si, P, S, Cl, Ar.
Oxides: Na_2O or Na_2O_2, MgO, Al_2O_3, SiO_2, P_4O_6 and P_4O_{10}, SO_2 and SO_3, Cl_2O or Cl_2O_7, none for Ar.
Chlorides: NaCl, $MgCl_2$, Al_2Cl_6, $SiCl_4$, PCl_3 and PCl_5, S_2Cl_2. (16)

2 Metallic: Na, Mg, Al. Ionic: Na_2O or Na_2O_2, MgO, Al_2O_3, NaCl, $MgCl_2$. Giant covalent lattice: Si, SiO_2. Simple covalent molecules: Si, P_4, S_8, Cl_2, P_4O_6 and P_4O_{10}, SO_2 and SO_3, Cl_2O or Cl_2O_7, Al_2Cl_6, $SiCl_4$, PCl_3, PCl_5, S_2Cl_2. (2)

3 $2Na(s) + O_2(g) \rightarrow 2Na_2O_2(s)$ (or $Na_2O(s)$).
$2Mg(s) + O_2(g) \rightarrow 2MgO(s)$.
$2Al(s) + 3O_2(g) \rightarrow 2Al_2O_3(s)$.
$4P(s) + 5O_2(g) \rightarrow P_4O_{10}(s)$.
$4P(s) + 3O_2 \rightarrow P_4O_6(s)$.
$S(s) + O_2(g) \rightarrow SO_2(g)$. (6)

4 $2Na(s) + Cl_2(g) \rightarrow 2NaCl(s)$.
$Mg(s) + Cl_2(g) \rightarrow MgCl_2(s)$.
$2Al(s) + 3Cl_2 \rightarrow Al_2Cl_6(s)$.
$Si(s) + 2Cl_2(g) \rightarrow SiCl_4(l)$.
$2P(s) + 3Cl_2(g) \rightarrow 2PCl_3(l)$.
$2S(s) + Cl_2(g) \rightarrow S_2Cl_2(l)$. (6)

5 $2Na + 2H_2O \rightarrow 2NaOH + H_2$, pH = 12–14.
$Mg + H_2O \rightarrow MgO + H_2$, pH = 7, but aqueous soln = 9–10.
$Cl_2 + H_2O = HCl + HClO$, pH = 3–4. (3)

6 The pH depends on the amount of water and other conditions, so there is a range of possible figures. Examiners will accept any figure in this range.
$Na_2O(aq) + H_2O(l) \rightarrow 2NaOH(aq)$, pH = 12–14.
$MgO(s) + H_2O(l) \rightarrow$ insoluble $Mg(OH)_2(s)$, pH = 9–10.
$Al_2O_3(s) + H_2O(l) \rightarrow$ does not dissolve in water, pH = 7.
$SiO_2(s) + H_2O(l) \rightarrow$ does not dissolve in water, pH = 7.
$P_4O_6(s) + 6H_2O(l) \rightarrow 4H_3PO_3(aq)$, pH = 2–5.
$P_4O_{10}(s) + 6H_2O(l) \rightarrow 4H_3PO_4(aq)$, pH = 1–3.
$SO_2(g) + H_2O(l) \rightarrow H_2SO_3(aq)$, pH = 3–5.
$SO_3(g) + H_2O(l) \rightarrow H_2SO_4(aq)$, pH = 1–3.
$Cl_2O(g) + H_2O(l) \rightarrow 2HClO(aq)$, pH = 3–5.
$Cl_2O_7^{2-}(s) + H_2O(l) \rightarrow 2HClO_4(aq)$, pH = 2–5. (10)

7 $NaCl(s) + (aq) \rightarrow Na^+(aq) + Cl^-(aq)$.
$MgCl_2(s) + (aq) \rightarrow Mg^{2+}(aq) + 2Cl^-(aq)$.
$Al_2Cl_6(s) + 6H_2O \rightarrow 2Al(OH)_3(s) + 6HCl(g)$.
$SiCl_4(l) + 4H_2O \rightarrow Si(OH)_4(s) + 4HCl(g)$.
$PCl_5(l) + 4H_2O \rightarrow H_3PO_4 + 5HCl(g)$. (5)

8 When water reacts with the simple covalent chloride $SiCl_4$, one of the water molecule lone pairs goes into the low-energy vacant orbitals in the $SiCl_4$ outer shell. The CCl_4 has no low-energy vacant orbitals in its outer shell so water does not react with it. In addition the C-Cl bonds are stronger than the Si-Cl bonds. (2)

(Total 50 marks)

CONCEPT TEST

1

Element	Mg	Al	Si	S
Oxide formula(e)	MgO	Al_2O_3	SiO_2	SO_2, SO_3
Bonding	I	I	C	C
Structure	GL	GL	GL	SM

(I = ionic, C = covalent, GL = giant lattice, SM = simple molecular.) (9)

a With oxygen, the silicon becomes covered by an unreactive oxide layer. (2)

b **i** $2Mg(s) + O_2(g) \rightarrow 2MgO(s)$.
ii $P_4O_6(s) + 6H_2O(l) \rightarrow 4H_3PO_3(aq)$.
iii $SO_2(g) + H_2O(l) \rightarrow H_2SO_3(aq)$. (3)

2

Element	Na	Al	Si	P
Chloride formula(e)	NaCl	Al_2Cl_6	$SiCl_4$	PCl_3, PCl_5
Bonding	I	C	C	C, C
Structure	GL	SM	SM	SM, SM

(I = ionic, C = covalent, GL = giant lattice, SM = simple molecular.) (9)

a PCl_3 is prepared by passing dry chlorine over heated red phosphorus. (3)

b **i** $2P(s) + 3Cl_2(g) \rightarrow 2PCl_3(l)$.
ii $2Na(s) + Cl_2(g) \rightarrow 2NaCl(s)$. (2)

c **i** $NaCl(s) + (aq) \rightarrow Na^+(aq) + Cl^-(aq)$.
ii $SiCl_4(l) + 4H_2O(l) \rightarrow Si(OH)_4(s) + 4HCl(aq)$. (2)

(Total 30 marks)

UNIT 7

RECALL TEST

1 Organic molecules that have the same functional group and react similarly form a homologous series. (1)

2 Methane CH_4, ethane C_2H_6, propane C_3H_8, butane C_4H_{10}, pentane C_5H_{12}, hexane C_6H_{14}, heptane C_7H_{16}, octane C_8H_{18}, nonane C_9H_{20}, decane $C_{10}H_{22}$. (3)

3 The strong covalent bonds in alkanes result in them being unreactive. The (C-C) and (C-H) bonds are strong (so have a high bond enthalpy – see unit 11), because the atoms are very small and unshielded, so the nuclei are held very strongly by the shared electrons. (1)

4 In the double bond the electrons in the pi bond are further from the nuclei than in a sigma bond, resulting in a weaker bond. Also the pi bond sticks out so is easily attacked by an electrophile. (1)

5 The nucleus of the large halogen atom is far from the shared electrons in the carbon–halogen bond. The electrons are also shielded by inner shells. (1)

6 **a** UV light or high temperature.

b Room temperature.

c Heat under reflux. (3)

7 a Structural isomers have the same molecular formula but different structural formulae. (1).

b Geometric isomers have the same molecular structure but differ in arrangement in space by having two different groups on the same side (*cis*) or on opposite sides (*trans*). (1)

8 5 (including cyclobutane and methylcyclopropane). (1)

9 9 structural and geometric isomers. (1)

10 a 4.

b 3.

c 2.

d 1. (4)

11 From C-F to C-I the bonds become weaker because the atoms are becoming larger so there is an increased distance between the nuclei and the bonding electrons, and there is increased shielding. (2)

12 (Induced Van der Waals = VdW, permanent dipole = PD, hydrogen bonds = H) alkanes VdW, alkenes VdW, halogenoalkanes PD, alcohols H, aldehydes PD, ketones PD, amines H, nitriles PD, carboxylic acids H, carboxylic salts ionic, esters PD, amides H, carbonyl chlorides PD. (7)

13

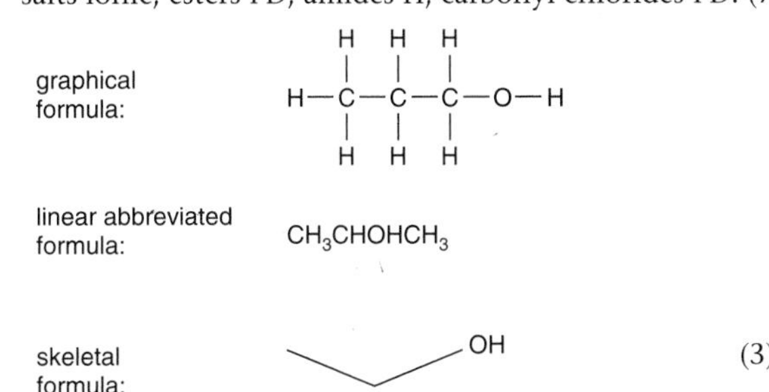

(3)

(Total 30 marks)

CONCEPT TEST

1 a $C_6H_{12} + Br_2 \rightarrow C_6H_{12}Br_2$. (1)

b The bonds in hexane are too strong to be broken by bromine, because the atoms in hexane are very small. UV light is required to initiate the reaction. (2)

2 a Room temperature.

b Heat under reflux.

c Ultraviolet light.

3 a The bonds become weaker because the atoms are becoming larger so there is an increase distance between the nuclei and the bonding electrons, and there is increased shielding. (3)

b The C=C bond is stronger than the C-C bond because there are more electrons involved in the C=C double bond. The C=C is less than double the strength of the single C-C covalent bond because both contain a single sigma bond, but the second bond in the C=C bonding is a pi bond in which the electrons are further from the nucleus than in a sigma bond. (3)

c The Si-Si bond is weaker because the Si atoms are larger so the electrons in the bond are further from the Si nucleus. (3)

4 a

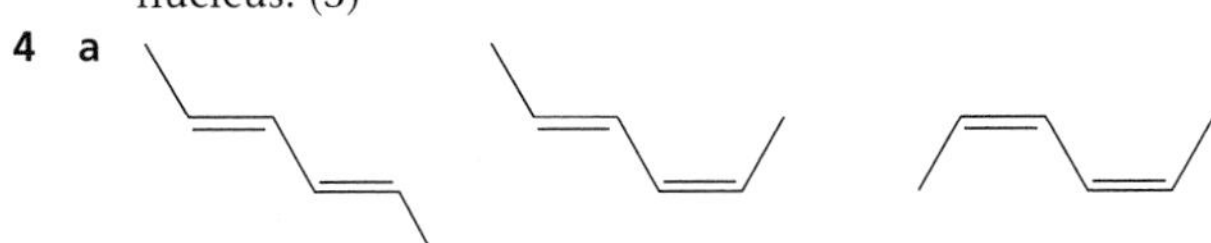

b The C=C make this molecule very reactive. (5)

(Total 20 marks)

UNIT 8

RECALL TEST

1 a Alkali.

b Acid.

c Oxidising agent.

d Reducing agent.

e Reagent that causes hydrolysis.

f Dehydrating agent. (6)

2 a Condensation/addition elimination. (1)

b Alkaline hydrolysis. (1)

3 a A molecule or ion with an electron-rich site which can donate a pair of electrons. (2)

b A molecule or ion with an electron-deficient site which can accept a pair of electrons. (2)

c When a group of atoms is added to a molecule, and no atoms are lost. (2)

d When an atom or one group of atoms is replaced by another group. (2)

4 a Nucleophilic substitution. (2)

b Electrophilic addition. (2)

(Total 20 marks)

CONCEPT TEST

1 a Oxidising agent: $KMnO_4$(aq) with H_2SO_4(aq)/$K_2Cr_2O_7$(aq) with H_2SO_4(aq).
Reducing agent: $LiAlH_4$(dry ether)/$NaBH_4$(aq)/H_2(g) with Ni(s) or Pt(s).
Dehydrating agent: concentrated H_2SO_4/solid P_4O_{10}. (3)

b When CH_3CH_2OH is converted to CH_2CH_2 then two H atoms and one O atom are lost: H_2O is lost, so the change is called dehydration. (1)

c Any sodium-containing base: NaOH(aq), or $NaHCO_3$(s), or Na(s). (1)

d Any strong acid: H_2SO_4(aq), or HCl(aq). (1)

2 a Condensation/addition elimination/esterification. (1).

b Water, as the reaction is hydrolysis. (1)

3 a A molecule or ion with an electron-rich site which can donate a pair of electrons replaces a group of atoms.(2)

b When bonds break and the pair of electrons go with one atom. (2)

c See Fig 8.5. You should draw the nucleophile as a CN^- ion rather than a OH^- ion. The lone pair is on the C in CN^-. (3)

4 a See Fig 8.6. Instead of the H-Br there is Br-Br, so a Br joins to the ethene first (instead of the H). (3)

b Bromine reacts by electrophilic addition, whereas the C=O reacts by nucleophilic addition. The Br_2 attacks the C=C to make Br-C-C^+. A C=O group cannot react in the same way, as the O atom cannot have a positive charge. Also, if a $C^{\oplus}$ formed that would mean a weak O-Br bond would have to form also. (2)

(Total 20 marks)

UNIT 9

RECALL TEST

1 Chlorine and ultraviolet light. (2)

2 More chlorine with ultraviolet light. (2)

3 A molecule or atom with an unpaired electron. (2).

4 Initiation: $Br_2 \rightarrow 2Br^\bullet$(g) with UV light.
Propagation: $CH_3CH_3 + Br^\bullet \rightarrow {}^\bullet CH_2CH_3$ + HBr (remember this one), then ${}^\bullet CH_2CH_3 + Br_2 \rightarrow CH_3CH_2Br + Br^\bullet$.
Termination: ${}^\bullet CH_2CH_3 + Br^\bullet \rightarrow CH_3CH_2Br$, *or* ${}^\bullet CH_2CH_3 + {}^\bullet CH_2CH_3 \rightarrow CH_3CH_2CH_2CH_3$ *or* on the reaction vessel walls $Br^\bullet + Br^\bullet \rightarrow Br_2$. (6)

5 By fractional distillation. (2)

6 To make useful alkenes and to turn cheap long-chain alkanes into short-chain alkanes which are worth more. (2)

7 Electrophilic addition. (1)

8 Room temperature. (1)

9 a $CH_2CH_2 + HBr \rightarrow CH_3CH_2Br$.

b $CH_2CH_2 + Br_2(CCl_4) \rightarrow CH_2BrCH_2Br$.

c $CH_2CH_2 + Br_2(aq) \rightarrow CH_2BrCH_2OH$ (mixed with CH_2BrCH_2Br).

d $CH_2CH_2 + H_2O \rightarrow CH_3CH_2OH$. (4)

10 $CH_2CH_2 + [O] + H_2O \rightarrow CH_2OHCH_2OH$. (1)

11 Hydrogen gas with nickel catalyst and 200 °C temperature. (3)

12 Alkenes decolorise bromine water. (1)

13 -[-$CH_2C(CH_3)H$-]-. (1)

14 Microbes cannot decompose them because they are chemically inert and waterproof. (2)

(Total 30 marks)

CONCEPT TEST

1 a Chlorine and ultraviolet light. (2)
b Heat under reflux with conc. sulphuric or phosphoric acids, then dilute with water, *or* pass ethene and steam over an acid catalyst. (2)
c Use dilute potassium permanganate, at room temperature. (2)
d Add hydrogen gas, Ni (or Pt) catalyst, 200 °C and high pressure. (2)

2 a Reagent: bromine water. (1)
Conditions: room temperature. (1)
Observation with palm oil: no change. (1)
Observation with sunflower oil: bromine water is decolorised. (1)
b Advantage: decomposes when discarded. Disadvantage: object could break down before it is finished with. (2)

3 a A molecule or atom with an unpaired electron. (2)
b The ultraviolet light in sunlight splits bonds to make free radicals. (1)
c Initiation: ultraviolet light + $Cl_2 \rightarrow 2Cl^\bullet$.
Propagation: $-CH_2- + Cl^\bullet \rightarrow -{}^\bullet CH- + HCl$, then $-{}^\bullet CH- + Cl_2 \rightarrow -CHCl- + Cl^\bullet$, or ${}^\bullet CH- + Cl^\bullet \rightarrow -CHCl-$. (3)

(Total 20 marks)

UNIT 10

RECALL TEST

1 Nucleophilic substitution. (2)

2 a $CH_3CH_2Cl(l) + NaOH(aq) \rightarrow CH_3CH_2OH(aq) + NaCl(aq)$.
b $CH_3CH_2I(l) + 2NH_3(ethanol) \rightarrow CH_3CH_2NH_2 + NH_4I$.
c $CH_3CH_2Br(l) + KCN(ethanol) \rightarrow CH_3CH_2CN + KBr$.
d $CH_3CH_2Br + KOH(ethanol) \rightarrow CH_2CH_2 + H_2O + KBr$. (8)

3 CH_3CH_2COOH, propanoic acid. (2)

4 Aqueous HCl, boil under reflux. (2)

5 Aqueous hydroxide ions with bromoethane produces ethanol. Hydroxide ions in ethanol produces ethene. (4)

6 Silver bromide, AgBr. It must be a bromoalkane. (2)

7 Short-chain alcohols hydrogen-bond with water, while long-chain alcohols are more like alkanes so held mostly by Van der Waals forces. (2)

8 Ethanol has hydrogen bonds, ethanal has only a permanent dipole. (2)

9 a Ethanol or hydroxyethane.
b Propan-2-ol.
c 2-methylpropan-2-ol.
d 2,2-dimethylpropan-1-ol. (8)

10 a $CH_3CH_2OH + Na \rightarrow CH_3CH_2O^-Na^+ + \frac{1}{2}H_2$.
b $CH_3CH_2OH + [O] \rightarrow CH_3COOH + H_2O$.
c $CH_3CH_2OH \rightarrow CH_2CH_2 + H_2O$.
d $CH_3COOH + CH_3CH_2OH \rightarrow CH_3COOCH_2CH_3 + H_2O$.
e $CH_3CH_2OH + PCl_5 \rightarrow CH_3CH_2Cl + POCl_3 + HCl$.
f $CH_3CH_2OH + HBr \rightarrow CH_3CH_2Br + H_2O$.
g $CH_3CHOHCH_3 + [O] \rightarrow CH_3COCH_3 + H_2O$. (8)

(Total 40 marks)

CONCEPT TEST

1 a $CH_3CH(CH_3)CH_2Br$.
b $CH_3CH_2CH(Br)CH_3$. (2)

2 a Aqueous NaOH produces propan-2-ol. (2)
NaOH in ethanol produces propene. (2)
b i NH_3(ethanol), heat under reflux. (2)
ii KCN(ethanol), heat under reflux.(2)
c Add aqueous silver nitrate. A cream precipitate forms if a C-Br group is present. (2)

3 a $CH_3CH_2CH_2CH_2OH$ or $CH_3CH(CH_3)CH_2OH$.
b $CH_3CH_2CHOHCH_3$, also written $CH_3CHOHCH_2CH_3$.
c $(CH_3)_3COH$. (3)

4 a A: ethanoic acid; conc. H_2SO_4; heat under reflux.
B: conc. HBr (or KBr and H_2SO_4); heat under reflux.
C: concentrated H_2SO_4; heat. (6)
b $CH_3COOCH(CH_3)_2$. (2)
c To help you: 2-methylpropan-2-ol is $(CH_3)_3COH$.
i Methylpropene $CH_2CH(CH_3)CH_3$.
ii It does not react. (2)

(Total 25 marks)

UNIT 11

RECALL TEST

1 Exothermic reactions have a negative enthalpy. (1)

2 a The enthalpy change when one mole of a substance is formed from its constituent elements in their normal states (under standard conditions). (3)
b The enthalpy change when one mole of a substance is completely combusted in oxygen (under standard conditions). (3)
c The enthalpy change when one mole of water is formed by reacting an acid and base (under standard conditions). (3)

3 The energy required to break one mole of a particular kind of bond in a particular compound. (3)

4 The average energy required to break one mole of a particular kind of bond taken from many compounds. (3)

5 The energy change accompanying a reaction is independent of the route taken. (2)

6 Quantity of heat (q) = mass (m) × specific heat capacity (c) × temperature change (ΔT). (2)

(Total 20 marks)

CONCEPT TEST

1 a +175 kJ mol^{-1}. (4)
b +35 kJ mol^{-1}. (4)
c ii −186 kJ mol^{-1}. (3)

2 9.7 kJ mol^{-1}. (2)

3 a −1169 kJ mol^{-1}. (3)
b 391 kJ mol^{-1}. (4)

(Total 20 marks)

UNIT 12

RECALL TEST

1 Lattice enthalpy is the energy change when one mole of an ionic solid is formed from its constituent gaseous ions. (1)

2 The ions in CaO both have a 2+ charge so the lattice energy would be very much greater than that for NaCl, as in NaCl the ions only have a 1+ charge. (1)

3 The Mg^{2+} is a much smaller ion than the Ca^{2+} ion. Mg^{2+} has one shell fewer than the Ca^{2+} ion. (1)

4 The Born–Haber diagram would be similar to Fig. 12.1, but would require the production of only one gaseous Cl^- ion, so either one $\Delta H_a(Cl)$ or half the (Cl-Cl) bond energy. (4)

5 Lattice energy is based on purely ionic compounds. Al_2Cl_6 is partially covalent. (1)

6 The diagram should show that
the sum of the hydration enthalpies
= lattice enthalpy + solution enthalpy. (2)

7 $Mg(OH)_2$ is insoluble because its ions are small so the lattice enthalpy is very large. $MgSO_4$ is soluble because the Mg^{2+} ion is small so the hydration enthalpy is large (while the sulphate anion is large so the lattice energy is low). (2)

8 From MgO to BaO the lattice energy decreases as the cationic radius increases, making the cation less attractive. (2)

9 With increasing atomic number the cations become larger so they polarise the anions less; thus the anions become less covalent, more ionic, and more stable. (2)

10 A strong acid is an acid that is fully dissociated (ionised). (1)

11 When strong acids and alkalis are dissolved in water they fully ionise, producing H^+ ions or OH^- ions. It is the reaction between these ions that produces heat when acids and alkalis are mixed. (1)

12 All the acid reacts with NaOH. As ethanoic acid is a weak acid some molecules of the acid are not ionised. When the alkali is added the ethanoic acid molecules first ionise, which costs energy, so overall the reaction is less exothermic. $CH_3COOH \rightarrow CH_3COO^- + H^+$. (2)

(Total 20 marks)

CONCEPT TEST

1 a The diagram should be similar to the $CaCl_2$ diagram (Fig. 12.1), except that the combined electron affinities are positive so the arrow should go up the diagram to the gaseous ions. CaO lattice enthalpy = $-3518\,kJ\,mol^{-1}$. (6)

b If you drew the diagram for $CaCl_3$ you would see that the 3rd ionisation energy was so large that it would make the ΔH formation massively endothermic. This means $CaCl_3$ would spontaneously turn into $CaCl_2$ and Cl_2. (2)

c Theoretical lattice energy is based on a purely ionic model. AgI is covalent so the experimental lattice enthalpy would be very different. (2)

d i $+1\,kJ\,mol^{-1}$. (2)

ii The energy released by the ions being surrounded by water supplies the energy required to separate the oppositely charged ions. (4)

2 $+45.6\,kJ\,mol^{-1}$. (5)

3 a $\Delta G = \Delta H - T\Delta S$. (1)

b i $-26.2\,kJ\,mol^{-1}$.

ii $14\,700\,J\,mol^{-1}$. (4)

iii The reaction is feasible at 1500 K as the ΔG is positive. It is not feasible at 500 K as the ΔG is negative. (2)

iv 1010 K (ΔG must be zero). (2)

(Total 30 marks)

UNIT 13

RECALL TEST

1 Rate is the measure of how fast a reactant (or product) concentration changes over time. The units of rate are usually $mol\,dm^{-3}\,s^{-1}$. (2)

2 Temperature, concentration, pressure, catalyst, surface area, light. (6)

3 Collisions with each other, with the correct orientation, and with enough energy. (3)

4 When the concentration is increased there are more molecules in the same volume, so they collide and react more often, and the rate is increased. (2)

5 When the surface area is increased there is more contact between the molecules so the rate increases. (2)

6 The diagram is like Fig. 13.3, but with the reactants lower than the products. Also the profile includes a low-energy state between two transition states. (2)

7 Activation energy is the minimum energy required in a molecular collision for the molecules to react. (3)

8 See Fig. 13.5. Did you ensure the areas under the curves are similar and that the higher temperature peak is lower than the lower temperature? Activation energy must be to the right of the peaks. (1)

9 When the temperature increases the molecular kinetic energy increases, so there are more molecules colliding with an energy greater than the activation energy. This is shown by the increase in the shaded area on the graph. (2)

10 A catalyst increases the reaction rate without being used up. (1)

11 A catalyst increases the reaction rate by creating an alternative reaction route with a lower activation energy. (2)

12 Catalysts produce an alternative reaction route by having a variable oxidation state, or by acting as a surface that adsorbs molecules, bringing them closer together. (2)

13 A kinetically stable substance does not react, because the activation energy is high so the rate is low. (1)

14 When a substance is thermodynamically stable then the reaction is not feasible. (1)

(Total 30 marks)

CONCEPT TEST

1 a The minimum energy required in a molecular collision for the molecules to react. (3)

b The diagram is like Fig. 13.3, but with the reactants lower than the products. Also the profile includes a low-energy state between two transition states. (3)

c See Fig. 13.5. (3)

d When the temperature increases, the molecular kinetic energy increases, so there are more molecules colliding with an energy greater than the activation energy. More molecules react, so the reaction rate increases. (3)

e Rate. (1)

2 a See Fig. 13.6. (2)

b A catalyst increases the reaction rate by creating an alternative reaction route with a lower activation energy so that there are more molecular collisions with an energy greater than the activation energy, so more molecules react. (3)

c White phosphorus is unreactive in water, so must be thermodynamically stable (assuming that it is not kinetically stable because it is so reactive in air). Phosphorus burns spontaneously in air so it must be thermodynamically unstable (it reacts) and kinetically unstable (it reacts quickly). (2)

(Total 20 marks)

UNIT 14

RECALL TEST

1 The rate equation links rate to the concentrations of the chemicals which control the reaction rate, e.g. rate = $k[HI]^2$. (1)

2 Order is the sum of the powers in the rate equation. (1)

3 $mol^{-1}\,dm^3\,s^{-1}$. (1)

4 s^{-1}. (1)

5 $mol\,dm^{-3}\,s^{-1}$. (1)

6 Order suggests the number of particles involved in the rate determining step. (1)

7 Half life is the time taken for the reactant concentration to halve. (1)

8 Colour, colorimeter; pH, pH meter; electrical conductivity, conductivity meter; polarised light, polarimeter; gas volume, gas syringe. (5)

9

Rate	[a]	[b]
1	1	1
2	2	1
3	3	1
4	1	2
9	1	3 (3)

10 See Fig. 14.1. (3)

11 The reaction is zero order. The gradient tells you that rate is constant. (2)

12 The reaction is first order. The gradient tells you the value for k. (2)

13 The reaction is second order. The gradient tells you the value for k. (2)

14 $E_a = 2830\,J\,mol^{-1}$. (1)

15 a Concentration against time and rate against concentration.

b Rate against concentration and half life against time.

c Rate against concentration squared. (5)

(Total 30 marks)

CONCEPT TEST

1 a Propanone = 1st order; I_2 = zero order; H^+ ions = 1st order. (3)

b I_2 is a reactant, but it is not in the rate determining step. (1)

c Rate = $k[CH_3COCH_3]^1\,[H^+]^1$. (1)

d Two. (1)

e The acid must be a catalyst as the acid is in the rate equation but is not used up by the reaction. (In fact, as H^+ ions are made by the reaction it is called an autocatalyst). (2)

f The rate determining step in a mechanism is the slowest step that controls the overall reaction rate. (2)
g CH_3COCH_3 and H^+ only. (Only they appear in the rate equation.) (2)

2 a You cannot write the rate equation as order is only determined experimentally. (2)
b *Either* by using a colorimeter which follows the $[MnO_4^-(aq)]$ *or* by using a pH meter which follows the $[H^+]$. (2)
c Half life is the time taken for the reactant concentration to halve. (2)
d X is zero order. (1)
e A graph of rate against [Y] should be a straight line if [Y] is first order. The graph will start at the origin. (1)

(Total 20 marks)

UNIT 15

RECALL TEST

1 'Dynamic equilibrium' is when in a reversible reaction the concentrations of reactants and products do not change, but the reactants are continually producing products and the products produce reactants. (1)

2 Le Chatelier's principle states that if the conditions of a system at equilibrium are changed then the equilibrium position will shift to resist the change. (1)

3 a Right.
b Left.
c Left.
d Does not change (as the reaction ΔH is zero).
e Does not change (as catalysts do not change the position of equilibrium). (5)

4 a Left.
b Left.
c Right.
d Does not change the pressure of the product. (4)

5

		forward	backward	yield
a	incr. temperature,	I	I	D
b	incr. pressure,	I	I	No
c	add catalyst.	I	I	No

(I = increase, D = decrease, No = no change.) (6)

6 a $N_2(g) + 3H_2(g) \rightleftharpoons 2NH_3(g)$.
Catalyst: iron. 450 °C and 200–1000 atm.
b $SO_2(g) + \frac{1}{2}O_2(g) \rightarrow SO_3(g)$.
Vanadium(V) oxide at 450 °C and 1–2 atm.
c $4NH_3(g) + 5O_2(g) \rightleftharpoons 4NO(g) + 6H_2O(g)$.
Pt/Rh catalyst at 850 °C and 1–2 atm. (3)

(Total 20 marks)

CONCEPT TEST

1 a If the temperature increased the yield would decrease, because an increase in temperature would shift the reaction position of equilibrium to the left as the forward reaction is exothermic. (2)
b High pressure would be very expensive, requiring thick-walled pipes and compressors. The yield is most economic at the stated temperature. (2)
c The catalyst would increase the rate of production of ammonia by obtaining the yield sooner as the rate increases. (2)
d The catalyst would not change the yield because catalysts do not influence yield, only rate. (2)

2 a 'Dynamic equilibrium' is when in a reversible reaction the concentrations of reactants and products do not change, but the reactants are continually producing products and the products produce reactants. (2)
b Increasing the temperature would shift the reaction equilibrium to the left, as the forward reaction is exothermic. (2)
c No, because there are more gas molecules on the product side of the reaction so an increase in pressure would shift the position of equilibrium to the side with fewer molecules (so resisting the increase in pressure slightly). (2)
d The yield is economic at low pressure. Higher pressure would be expensive and would decrease the yield. (2)
e $2NO_2(g) + H_2O(l) + \frac{1}{2}O_2(g) \rightarrow 2HNO_3(aq)$ (Other equations are possible. In some plants excess NO_2 is distilled off). (2)
f Vanadium pentoxide (V_2O_5), 450 °C, 1–2 atm. (2)

(Total 20 marks)

UNIT 16

RECALL TEST

1 Kinetics, equilibrium, enthalpy, economic, and environmental factors. (5)
2 Temperature, concentration, pressure, catalysts, surface area (or light). (5)
3 Temperature, concentration, pressure, and economic factors. (4)
4 Low. (1)
5 Both increase costs unless near-atmospheric conditions. (1)
6 $SO_2(g) + \frac{1}{2}O_2(g) \rightleftharpoons SO_3(g)$. (1)
7 Both yield and rate would decrease. (1)
8 Increased yield. Decreased rate. (1)
9 Increased rate. No effect on yield. (1)
10 1–2 atm pressure, 450 °C, and V_2O_5 catalyst. (2)
11 SO_3 is dissolved in pure H_2SO_4 which is then diluted. (1)
12 $4NH_3(g) + 5O_2(g) \rightleftharpoons 4NO(g) + 6H_2O(g)$. (1)
13 $2NO(g) + O_2(g) \rightarrow 2NO_2(g)$. (1)
14 Fertilisers, explosives, and polyamides. (3)
15 Ammonium sulphate/ammonium nitrate. (1)
16 Lead compounds were added to limit pre-ignition. (1)
17 Lead harms the nervous system, and poisons catalytic converters. (2)
18 They convert pollutants (CO, NO_x, unburnt hydrocarbons) into CO_2, N_2, and H_2O. (3)
19 Rhodium, Rh. (1)
20 Catalytic poisons bind irreversibly with catalysts. (1)
21 Ethanol is made by hydrating ethene over a catalyst of phosphoric acid at 300 °C and 70 atm. (1)
22 Alcoholic drinks, fuels, solvents. (1)
23 Petrol additive and as an industrial feed stock. (1)

(Total 40 marks)

CONCEPT TEST

1 a Pre-ignition is when fuel combusts too early in a car engine. It causes damage to the car engine, loss of power, and increased fuel consumption. (2)
b Otherwise methanol production would be too slow to be economic. (2)
c Yield of methanol would increase/equilibrium would shift to right. The rate would increase. (2)
d Decrease the yield/equilibrium shifts to the left (exothermic reaction). Increase the rate. (2)
e Improved combustion so less CO, NO_x, unburnt fuel. Less sulphur dioxide/acid rain. (2)
f Lessen NO_x emissions, so less acid rain. Less poisonous CO and unburnt fuel. (3)
g Heterogeneous, because the solid catalyst is in a different phase to the exhaust gases. (2)

(Total 15 marks)

UNIT 17

RECALL TEST

1 If the conditions of a system at equilibrium are changed then the equilibrium will shift to resist the change. (2)
2 $mol\,dm^{-3}$. (1)
3 a $\dfrac{[SO_3(g)]_{eqm}^2}{[SO_2(g)]_{eqm}^2 \times [O_2(g)]_{eqm}}$.
b $mol^{-1}\,dm^3$. (2)
4 particular/constant. (1)
5 a $K_c = [CO_2(g)]_{eqm}$ (K_c so square brackets are required). (1)
b $mol\,dm^{-3}$. (1)

6 Moles glucose = 0.0278, moles water = 5.5556, mole fraction water = 0.995. (3)
7 Partial pressure = mole fraction of a gas multiplied by the total pressure. (1)
8 Moles O_2 = 0.875, moles He = 12/4 = 3, partial pressure oxygen = 11.3. (3)
9 Total pressure = 1 + 10 + 80 = 91 kPa. (1)
10 a $K_p = \frac{(pSO_3)^2}{pO_2 \times (pSO_2)^2}$. (1)
b Pa^{-1}. (1)
11 K_c and K_p vary with temperature. (1)
12 Temperature. (1)

(Total 20 marks)

CONCEPT TEST

1 a $K_c = \frac{[CH_3COOCH_2CH_3]\,[H_2O]}{[CH_3COOH]\,[CH_3CH_2OH]}$

$K_c = \frac{0.2 \times 0.2}{0.8 \times 0.8} = 0.0625$ (no units)

(No units as vols cancel.) (6)
b Concentrated sulphuric acid absorbs water, so the amount of water will be decreased, so the equilibrium position will shift to produce more water and so increase the ester yield. (3)
c Catalysts do not change the equilibrium position so the yield is unchanged. (2)
2 a The reaction is endothermic, so an increase in temperature shifts the equilibrium position to the right, so increasing the yield. (2)
b $K_p = \frac{pCO(g) \times (pH_2(g))^3}{pCH_4(g) \times pH_2O(g)}$. (1)
c Pa^2. (1)
d Pressure does not change the value of an equilibrium constant. (2)
3 a Partial pressure of HI = 7.5 kPa. (3)
b $K_p = \frac{pH_2(g) \times pI_2(g)}{(pHI(g))^2}$
= 0.0078 (no units.) (5)
c The amount decomposed will not change as there are the same number of gaseous molecules on either side of the reaction. (3)
d i $K_p = pCO_2(g)$. (1)
ii The amount of solid does not influence the position of equilibrium, so the partial pressure $pCO_2(g)$ will not change. (1)

(Total 30 marks)

UNIT 18

RECALL TEST

1 a proton donor, a proton acceptor. (2)
2 fully ionized. (1)
3 a $pH = -\log_{10} [H^+(aq)]$. (1)
b $pK_a = -\log K_a$. (1)
c $pK_w = -\log K_w$. (1)
d $pOH = -\log [OH^-(aq)]$. (1)
4 a pH = 1.3.
b 1.3.
c pH = 1.
d 2.8. (4)
5 $HCOOH(aq) \rightleftharpoons HCOO^-(aq) + H^+(aq)$. (1)
6 pH = 12.7. (1)
7 A buffer solution is a chemical mixture that resists changes in pH on addition of small amounts of acid or alkali. (Don't say it stops the pH changing, as the pH does change slightly). (2)
8 pH = 4.76. (1)
9 pH = 5.0. (2)
10 See Fig 18.3 for answers. (2)

(Total 20 marks)

CONCEPT TEST

1 a i proton acceptor.
ii $pH = -\log [H^+(aq)]$, where $[H^+]$ is H^+ ion concentration in mol dm^{-3}.
iii $pK_w = -\log K_w$. (3)
b pH = 1.7. (2)
c pH = 2.26. (3)
d pH = 6.76. (2)
e pH = 10.5. (2)
2 a Acid added: $H^+(aq) + CO_3^{2-}(aq) \rightarrow HCO_3^-(aq)$.
Alkali added: $HCO_3^-(aq) + OH^-(aq) \rightarrow CO_3^{2-}(aq) + H_2O(l)$. (2)
b i $NaHCO_3 + NaOH \rightarrow Na_2CO_3 + H_2O$.
The Na_2CO_3 dissolves to form CO_3^{2-} ions. $[CO_3^{2-}(aq)] = 0.05$ mol dm^{-3}. (2)
ii The $NaHCO_3(aq)$ left over forms the HCO_3^- ions. So $[HCO_3^-] = 0.05$ mol dm^{-3}. (2)
iii As the $[CO_3^{2-}(aq)] = [HCO_3^-]$, so pH = pK_a, so pH = 10.3. (2)

(Total 20 marks)

UNIT 19

RECALL TEST

1 The p orbitals on each carbon atom overlap with the adjacent p orbitals to merge into one delocalised ring of pi electrons. (1)
2 The enthalpy of delocalisation may be determined by comparing the benzene hydrogenation with the hydrogenation of a simple alkene (e.g. cyclohexene). The benzene hydrogenation will be found to be less than three times the cyclohexene hydrogenation. The difference is due to the enthalpy of delocalisation. An energy-level diagram would illustrate the ideas effectively. (3)
3 Same as Fig. 19.5. (2)
4 Same as Fig. 19.6. (2)
5 a $C_6H_6 + HNO_3 \rightarrow C_6H_5NO_2 + H_2O$.
b $C_6H_6 + Cl_2 \rightarrow C_6H_5Cl + HCl$.
c $C_6H_6 + Br_2 \rightarrow C_6H_5Br + HBr$.
d $C_6H_6 + CH_3COCl \rightarrow C_6H_5COCH_3 + HCl$. (4)
6 In hydrated aluminium chloride the aluminium atoms would be surrounded by water molecules, so lone pairs from the Cl would not be able to join with the Al. The Al may no longer act as a halogen carrier. (2)
7 Benzene with ethene and HCl. (2)
8 Explosives and dyes. (2)
9 Benzoic acid. (1)
10 a Aqueous potassium permanganate with sulphuric acid and heat.
b Tin and concentrated hydrochloric acid and heat.
c Hydrochloric acid (or sulphuric acid) and sodium nitrite, at 5 °C.
d Phenol (or 2-naphthol) with aqueous NaOH.
e Aqueous NaOH.
f Aqueous dilute nitric acid. (6)

(Total 25 marks)

CONCEPT TEST

1 a Same as Fig. 19.6. (5)
b Electrophilic substitution. (1)
2 a B is $CH_3C_6H_4NO_2$. (2)
b Tin and concentrated hydrochloric acid. (2)
c $CH_3C_6H_4N_2^+$. (1)
d An azo dye. (1)
e You should have drawn an azo dye. The N=N link in the middle contains two bonds. Similar to Fig. 19.10, but with a methyl group on one benzene ring and the OH^- group on the other. (1)
f With iron(III) chloride solution phenol will produce a violet colour. (2)
3 a Chloromethane and aluminium chloride, heat.
b Chlorine and UV light.

c Aqueous sodium hydroxide.
d Aqueous potassium permanganate(VII) and warm.
e Aqueous sodium hydroxide, room temperature. (5)

(Total 20 marks)

UNIT 20

RECALL TEST

1 Dipole–dipole forces (and Van der Waals forces). (1)
2 nucleophilic addition, alcohol, aldehydes, carboxylic acid. (5)
3 Similar to Fig. 20.3, but propanone replaces the ethanal molecule. (3)
4 a If the pH is too low, many H^+ ions join the CN^- ion to make HCN, lowering the $[CN^-]$. (1)
b If the pH is too high then the second stage of the mechanism is slow. (1)
5 2-hydroxypropanenitrile. (1)
6 Reagent is 2,4-dinitrophenylhydrazine; produces a brightly coloured solid. (2)
7 a Same as Fig. 20.5. (1)
b Same as Fig. 20.6.
Name: ethanal 2,4-dinitrophenylhydrazone. (2)
8 a Reduction, propan-1-ol.
b Reduction, propan-2-ol. (2)
9 a $CH_3CH_2CHO + 2[H] \rightarrow CH_3CH_2CH_2OH$.
b $CH_3COCH_3 + 2[H] \rightarrow CH_3CHOHCH_3$.
c $CH_3CHO + HCN \rightarrow CH_3CH(OH)CN$.
d $CH_3CHO + (O) \rightarrow CH_3COOH$. (4)
10 Fehling's and Tollen's reagents. (2)
11 Propanone to A: HCN(aq) at pH = 5 (or + KCN).
A to B: aqueous HCl and heat under reflux. (2)
12 a Aqueous HCl and heat under reflux.
b The reducing agent $NaBH_4$(aq) is used, or $LiAlH_4$ (dry ether).
c KCN(ethanol) and heat under reflux. (3)

(Total 30 marks)

CONCEPT TEST

1 a X is propanone, CH_3COCH_3. Draw all bonds. (1)
b Draw diagram similar to Fig. 20.3. Check that the arrow goes from the C in CN^-, and that the CN^- is the correct way around. (3)
c In the second step of the mechanism in **b** H^+ ions are necessary. If the pH is high, then the H^+ concentration will be low, so the rate of this second step will be slow. (2)
d Aqueous HCl and heat under reflux. (2)
e Same as Fig. 20.5. (1)
f *Either* add Fehling's which will produce a red solid, *or* add Tollen's reagent which will produce a silver mirror (or black solid). (2)
g Recrystallisation. (1)
h The melting point of the pure derivative, X, could be determined and the reading checked with figures in a data book. (2)
2 a $NaBH_4$ is a reducing agent. (1)
b i $CH_3CH_2CH_2OH$.
ii $CH_3CHOHCH_3$.
iii $CH_3CH_2CH_2NH_2$. (3)
c i $CH_3CHO + 2[H] \rightarrow CH_3CH_2OH$.
ii $CH_3CHO + (O) \rightarrow CH_3COOH$. (2)

(Total 20 marks)

UNIT 21

RECALL TEST

1 a $CH_3COOH + NaOH \rightarrow CH_3COO^-Na^+ + H_2O$.
b $CH_3COOH + NaHCO_3 \rightarrow CH_3COO^-Na^+ + H_2O + CO_2$.
c $CH_3COOH + PCl_5 \rightarrow CH_3COCl + POCl_3 + HCl$.
d $CH_3COOH + CH_3CH_2OH \rightarrow CH_3COOCH_2CH_3 + H_2O$.
e $CH_3COOH + 4[H] \rightarrow CH_3CH_2OH + H_2O$. (5)
2 Blue damp litmus turns pink/red, $NaHCO_3$ gives off CO_2 bubbles, PCl_5 fizzes/produces steamy fumes, ethanol with concentrated H_2SO_4 produces a smell (often fruity). (4)
3 Propanoic acid, ethanol, and conc. H_2SO_4. (2)
4 $CH_3COOCH_2CH_3 + H_2O \rightarrow CH_3COOH + CH_3CH_2OH$. (1)
5 Fat boiled in aqueous alkali makes soap. (1)
6 a $CH_3COCl + H_2O \rightarrow CH_3COOH + HCl$.
b $CH_3COCl + CH_3CH_2OH \rightarrow CH_3COOCH_2CH_3 + HCl$.
c $CH_3COCl + 2NH_3 \rightarrow CH_3CONH_2 + NH_4Cl$.
d $CH_3COCl + CH_3CH_2NH_2 \rightarrow CH_3CONHCH_2CH_3 + HCl$.
e $CH_3NH_2 + HCl \rightarrow CH_3NH_3^+ + Cl^-$.
f $NH_3 + CH_3Br \rightarrow HBr + CH_3NH_2$.
g $CH_3NH_2 + CH_3Br \rightarrow HBr + (CH_3)_2NH$. (7)
7 a Dry ethanol at room temperature.
b Aminomethane at room temperature.
c Aqueous NaOH.
d PCl_5 at room temperature. (8)
8 a H_2SO_4(aq) or NaOH(aq), and heat under reflux.
b NaOH(aq) and heat under reflux. (2)

(Total 30 marks)

CONCEPT TEST

1 a P: CH_3COOH; Q: CH_3COCl; R: $CH_3COOCH_2CH_3$; S: $CH_3COO^-Na^+$; T: CH_3CH_2OH; U: CH_3NH_2; V: $CH_3CONHCH_3$. (7)
b i $CH_3COOH + NaOH \rightarrow CH_3COO^-Na^+ + H_2O$ (or use $NaHCO_3$).
ii $CH_3COCl + CH_3CH_2OH \rightarrow CH_3COOCH_2CH_3 + HCl$.
iii $CH_3NH_2 + CH_3COCl \rightarrow CH_3CONHCH_3 + HCl$ *or* $2CH_3NH_2 + CH_3COCl \rightarrow CH_3CONHCH_3 + CH_3NH_3^+Cl^-$.
iv $CH_3COOCH_2CH_3 + NaOH \rightarrow CH_3COO^-Na^+ + CH_3CH_2OH$. (4)
c i Room temperature.
ii Room temperature.
iii Heat under reflux. (6)
2 a Similar to Fig. 8.5, but the nucleophile is CH_3NH_2, so the arrow goes from the N. (4)
b i Dry/room temperature.
ii N-methylethanamide.
iii Substituted amide. (3)
c i *Either* add water – effervescence will only occur with the CH_3COCl, *or* add $AgNO_3$(aq), when only the CH_3COCl would form a white precipitate, *or* add PCl_5 – white fumes will be vigorously produced only with CH_3COOH.
ii *Either* add blue litmus which turns pink to blue only with CH_3COOH, *or* add $NaHCO_3$ which bubbles only with CH_3COOH, *or* add PCl_5 – white fumes will be vigorously produced only with the CH_3COOH. (6)

(Total 30 marks)

UNIT 22

RECALL TEST

1 $HOOC(C_6H_4)COOH$ and ethane-1,2-diol, $HOCH_2CH_2OH$. (2)
2 a $[-OOC(C_6H_4)COOCH_2CH_2O-]_n$.
b $[-HN(CH_2)_6NHOC(CH_2)_8COO-]_n$.
c $[-OCH(CH_3)CH_2COO-]_n$.
d $[-HNC(R)HCOO-]_n$. (4)
3 Any triol, e.g. $C(CH_2OH)_3$. (1)
4 a $H_2N(CH_2)_6NH_2$ and $HOOC(CH_2)_4COOH$.
b $H_2N(C_6H_4)NH_2$ and $HOOC(CH_2)_4COOH$.
c $H_2NC(R)HCOOH$. (3)
5 a Terylene is used to make fibres.
b Nylon 6,6 is used as a fibre in clothes, in ropes to tie up ships, and hard-wearing parts of home appliances.
c Kevlar is used in bullet-proof vests and to protect tyres and canoes. (3)
6 hydrolyse. (1)
7 a $^+H_3NCH(R)COO^-$.
b $^+H_3NCH(R)COOH$.
c $H_2NCH(R)COO^-$. (3)
8 a Addition.
b Polypeptide.
c Polyester.

d Polyamide.

e Polyester. (5)

9 Optical isomerism occurs when two isomers have the same structure, but have a different arrangement in space by being mirror images of each other that cannot be superimposed. (2)

10 * indicates the chiral carbon atoms: $HOCH_2C^*HClCH_2C^*H(CH_3)COOH$. (2)

11 Optical isomers rotate the plane of polarised monochromatic light in opposite directions. To measure this a polarimeter may be used. (2)

12 You must draw clear 3D diagrams. The bonds should suggest the 109° bond angle or you lose a mark. Similar to Fig. 22.5. (2)

(Total 30 marks)

CONCEPT TEST

1 a 1: $H_2N(CH_2)_6NH_2$ and $HOOC(CH_2)_4COOH$.
2: $H_2NC(CH_3)HCOOH$, $H_2NC(CH_2CH_3)HCOOH$.
3: $HOOC(C_6H_4)COOH$ and $HOCH_2CH_2OH$.
4: CH_2CH_2. (7)

b 1: polyamide/nylon; 2: polypeptide; 3: polyester; 4: polyalkane. (4)

c 1: hydrogen bonding; 2: hydrogen bonding; 3: dipole–dipole; 4: Van der Waals forces. (2)

2 a i $[\text{-CO}(C_6H_4)\text{CONH}(C_6H_4)\text{NH-}]_n$.
ii $[\text{-HN}(CH_2)_5\text{COO-}]$
iii $[\text{-OOC}(C_6H_4)\text{COOCH}_2\text{CH}_2\text{O-}]_n$. (6)

b Condensation/addition elimination. (1)

(Total 20 marks)

UNIT 23

RECALL TEST

1 a Bromine water turns colourless. Purple $KMnO_4$ goes colourless. (2)

b Purple $KMnO_4$ goes colourless. Orange $K_2Cr_2O_7$ goes green. Addition of conc. ethanoic acid (and $H_2SO_4(l)$) produces a sweet smell. PCl_5 vigorously produces white fumes. (4)

c Blue litmus turns pink/red. Any carbonate (e.g. $NaHCO_3(s)$) produces effervescence. Addition of ethanol (and $H_2SO_4(l)$) produces a sweet smell. PCl_5 vigorously produces white (HCl) fumes. (4)

d 2,4-dinitrophenylhydrazine produces a brightly coloured solid. (1)

e Fehling's produces a red solid. Tollen's produces a silver mirror/black precipitate. (2)

f Iodoform test. Either iodine with NaOH(aq), or sodium chlorate(I) with KI. The group produces a pale precipitate. (1)

g Add aqueous NaOH, warm gently, neutralise with dilute nitric acid. Add $AgNO_3(aq)$ and shake. A cream precipitate forms. (1)

h $FeCl_3$ forms a violet solution. Phenol will turn blue litmus pink/red. (2)

2 a Blue flame. (1)

b Yellow, very sooty flame. (1)

3 bond vibrations. (1)

(Total 20 marks)

CONCEPT TEST

1 a i P and R; brown bromine water is decolorised, detecting an alkene group >C=C<. (3)

ii Q and R; produces a violet colour, detecting a phenolic group/an OH group directly attached to an aromatic ring. (3)

iii P only; turns from orange to green, detects the aldehyde group -CHO. (3)

b P. On the IR spectrum there is absorbance at 1680 and 1630 cm^{-1}, indicating C=C, an alkene group. Also there is absorbance at 1130 cm^{-1} due to the aldehyde group, -CHO. Only P has these groups. (3)

2 a 122 (largest mass/charge ratio). (1)

b -COOH, carbonylic acid, because there is absorbance in the range 2500–3300 cm^{-1}. It is not an alcohol OH group as the chart lacks absorbance at 3200–3550 cm^{-1}. (4)

c X must be benzoic acid C_6H_5OOH. (1)

d The chemical shifts would be at between 7.3 and 9.7. (2)

(Total 20 marks)

UNIT 24

RECALL TEST

1 An element whose atom's last electron went into a d subshell is called a d-block element. (1)

2 4s. (1)

3 a Mn: $1s^2\ 2s^2\ 2p^6\ 3s^2\ 3p^6\ 3d^5\ 4s^2$.

b Fe^{3+}: $1s^2\ 2s^2\ 2p^6\ 3s^2\ 3p^6\ 3d^5\ 4s^2$. (2)

4 A stable $3d^5$ is formed when Fe^{2+} oxidises to Fe^{3+}, but it is difficult to oxidise Mn^{2+} to Mn^{3+} because it means losing an electron from the stable $3d^5$. (3)

5 Nearby ligands split the d subshell energy levels. Certain colours (frequencies) are absorbed by electrons which are promoted from a lower d orbital to a higher d orbital. The split in energy levels is equivalent to visible (and ultraviolet) parts of the spectrum. (4)

6 Element present, oxidation number, ligands present. (3)

7 Complexes are formed when a central cation (or atom) accepts dative covalent (co-ordinate) bonds from ions (or molecules). (2)

8 Ligands have lone pairs which form dative bonds with a central cation or atom in a complex. (2)

9 A ligand that uses one lone pair per molecule in complexes. (1)

10 Aqua H_2O, hydroxo OH^-, ammine NH_3, chloro Cl^-, oxo O^{2-}, cyano CN^-, thiocyano SCN^-, carbonyl CO. (6)

11 Same as Fig. 24.6. (2)

12 6. (1)

13 The transition metal atoms' successive ionisation energies gradually increase, and the energy required for the higher oxidation numbers is compensated by the covalent bond formation (e.g. in MnO_4^-) or high lattice energy or hydration energy. (2)

(Total 30 marks)

CONCEPT TEST

1 a A transition element is an element that forms at least one ion with a partially filled d subshell. (2)

b Sc and Zn. (1)

c See answer to Recall Test question **5** above. (4)

d The transition metal atoms' successive ionisation energies gradually increase. (2)

e Covalent bond formation (e.g. in MnO_4^-)/or high lattice energy/hydration energy. (1)

2 a See answer to Recall Test question **8** above. (2)

b i Oxo vanadium(IV) ions.
ii Dichlorocopper(I) ions (or dichlorocuprate(I) ions).
iii Tetracarbonyl nickel. (6)

c

		3d					4s	
Fe atom	(Ar)	[↑↓]	[↑]	[↑]	[↑]	[↑]	[↑↓]	
Fe^{2+} ion	(Ar)	[↑↓]	[↑]	[↑]	[↑]	[↑]	[ ]	
Fe^{3+} ion	(Ar)	[↑]	[↑]	[↑]	[↑]	[↑]	[ ]	(3)

d A stable $3d^5$ is formed when Fe^{2+} oxidises to Fe^{3+}, but it is difficult to oxidise Mn^{2+} to Mn^{3+} because it means breaking into the particularly stable $3d^5$. (4)

3 a *Either* diaminoethane, $NH_2CH_2CH_2NH_2$, *or* ethanedioate ions (oxalate ions) $C_2O_4^{2-}$. (2)

b 3. (1)

c Optical isomerism. (1)

d Octahedral. (1)

(Total 30 marks)

UNIT 25

RECALL TEST

1 Cr^{3+} green, Mn^{2+} colourless, Fe^{2+} blue-green, acidified Fe^{3+} yellow, Co^{2+} pink, Ni^{2+} green, Cu^{2+} blue, Zn^{2+} colourless. (8)

2 **a** Redox.
b Deprotonation.
c Deprotonation.
d Ligand exchange. (4)

3 Cr^{3+} YYYNN, Mn^{2+} YNYNY, Fe^{2+} YNYNY, Fe^{3+} YNYNN, Co^{2+} YNYYY, Ni^{2+} YNYYN, Cu^{2+} YNYYN, Zn^{2+} YYYYN. (Y = Yes, N = No.) (8)

4 **a** $4Fe(OH)_2 + O_2 + 2H_2O \rightarrow 4Fe(OH)_3$.
b $[Cr(OH)_3(H_2O)_3] + 3OH^- \rightarrow [Cr(OH)_6]^{3-}$.
c $[Cr(OH)_3(H_2O)_3] + 3H^+ \rightarrow [Cr(H_2O)_6]^{3+}$.
d $2Cr^{3+} + 3H_2O_2 + 10OH^- \rightarrow 2CrO_4^{2-} + 8H_2O$. (4)

5 V +5 is reduced to V +4 by dilute H_2SO_3 (sulphuric(IV) acid), made from sodium sulphite and hydrochloric acid; V +4 is reduced to V +3 by hot H_2SO_3 (aq), or by cold powdered Zn. V +3 is reduced to V +2 by boiling-hot Zn in HCl(aq). Or all: reduction by Zn(s) with HCl(aq). (5)

6 Aqueous $KMnO_4$ with H_2SO_4. (1)

(Total 30 marks)

CONCEPT TEST

1 **a** i $Fe(OH)_2$. (1)
ii Brown/red. (1)
iii $4Fe(OH)_2(s) + O_2(g) + 2H_2O(l) \rightarrow 4Fe(OH)_3(s)$. (2)
b i $Fe^{2+} \rightarrow Fe^{3+} + e^-$. (2)
ii $MnO_4^-(aq) + 8H^+(aq) + 5e^- \rightarrow Mn^{2+}(aq) + 4H_2O(l)$. (2)
iii $MnO_4^-(aq) + 5Fe^{2+}(aq) + 8H^+(aq) \rightarrow Mn^{2+}(aq) + 5Fe^{3+}(aq) + 4H_2O(l)$. (2)

2 **a** Deprotonation/acid–base. (1)
b Ligand exchange. (1)
c By the addition of aqueous acid, e.g. dilute sulphuric acid. (2)
d By the addition of conc. HCl. (2)

3 **a** Zn dust in HCl(aq).
b $KMnO_4$ with $H_2SO_4(aq)$, H_2O_2 in NaOH(aq).
c Any acid.
d Aqueous NaOH. (4)

(Total 20 marks)

UNIT 26

RECALL TEST

1 The potential difference between a half cell and the standard hydrogen electrode under standard conditions when no current flows. (3)

2 Same as Fig. 26.1. (7)

3 Ag is positive electrode. Feasible reaction: $Ni(s) + 2Ag^+(aq) \rightarrow Ni^{2+}(aq) + 2Ag(s)$.
e.m.f. = 0.33 volts. (3)

4 **a** $Pt(s) \mid H_2(g), H^+(aq) \,¦\, Ag^+(s) \mid Ag(s)$.
b $Pt(s) \mid Fe^{2+}(aq)/Fe^{3+}(aq) \,¦\, Ni^{2+}(aq) \mid Ni(s)$. (4)

5 It must be due to the activation energy being high, so rate low. (2)

6 If non-standard conditions are being used. (2)

7 water, oxygen. (2)

8 methane/hydrogen, a hydrocarbon, oxygen/air. (3)

9 Fuel cells are much more energy efficient, more mobile and resist damage. (3)

10 electrochemical series. (1)

(Total 30 marks)

CONCEPT TEST

1 **a** For oxygen to oxidise a halogen the oxygen $E^\ominus$ must be more positive than the halogen $E^\ominus$. The equation with the more negative $E^\ominus$ is where oxidation occurs. From the table only bromide and iodide are oxidised by oxygen under standard conditions. (2)
b The reaction may be feasible but the activation energy must be too high, so the rate is low. (2)

2 **a** Iron(III) ions will oxidise the iodide ions, because the $Fe^{3+}(aq)/Fe^{2+}(aq)$ $E^\ominus$ is more positive than the iodine/iodide $E^\ominus$. Oxidation occurs at the more negative $E^\ominus$. (2)
b The $E^\ominus$ indicate that this reaction does not occur under standard conditions, so non-standard conditions must be being used, e.g. the iodine and iron(II) ion concentrations may be more than 1 mol dm^{-3}. (2)
c The $Fe^{3+}(aq)/Fe^{2+}(aq)$ $E^\ominus$ is more negative than F, G, H, so they will oxidise iron(II) ions to form iron(III) ions. (2)
d For oxygen to be reduced (gain electrons) the oxygen-containing $E^\ominus$ must be more positive than the iron-containing $E^\ominus$, so the iron-containing species that would be oxidised by oxygen are metallic iron, iron(II) ions, so the iron ends up oxidised to iron(III). (2)
e $Fe(s) \mid Fe^{2+}(aq) \,¦\, O_2(g), OH^-(aq) \mid Pt(s)$. (2)
f Zinc/zinc(II) ion $E^\ominus$ is more negative than iron $E^\ominus$ so the zinc will oxidise (corrode) in preference to the iron. (2)
g Nickel will protect the iron from water and oxygen which is necessary for rusting, until the nickel is scratched, because then the oxygen and water could make contact with the iron, starting rusting. (2)
h Disproportionation is the simultaneous oxidation and reduction of the same element, so the element must start at an intermediate oxidation number and then be oxidised and reduced. Fe +2 could be oxidised to Fe +3 and reduced to iron, but the $E^\ominus$ show that the opposite occurs. Fe +3 is another intermediate oxidation number, but the $E^\ominus$ indicate that Fe +3 would not disproportionate because the 2nd $E^\ominus$ is more negative than the 3rd. (2)

(Total 20 marks)

UNIT 27

SYNOPTIC EXAM-STYLE QUESTIONS

1 **a** $Ag^+(aq) + Cl^-(aq) \rightarrow AgCl(s)$. (1)
b 87.61, so X is strontium. (4)
c By adding chlorine water. This produces a brown iodine with iodide. To confirm the presence of iodine, *either* add starch which turn blue-black *or* add hexane, which produces a purple layer. (2)
d Lattice energy is for purely ionic compounds. Iron(III) chloride must have some covalent character. (2)
e $CuCl_4^-$. The different ligands change the energy-level gap within the 3d subshell. It is this energy-level gap which absorbs particular colours. If the gap changes then the colour absorbed changes, so the substance appears a different colour. (3)
f i By reduction and oxidation/electron transfer.
ii By nucleophilic substitution.
iii By elimination.
iv By free-radical substitution.
v By electrophilic substitution. (5)
g The hydroxide ions are nucleophiles (have a lone pair) so they will substitute the chloride, by nucleophilic substitution. The Cl on the benzene ring,$ClC_6H_4CH_3$, is protected from the nucleophilic attack by the delocalised pi bonding which would repel the incoming lone pair. (3)

(Total 20 marks)

2 **a** –724 kJ mol^{-1}. (3)
b i The equilibrium will shift to the left because the reaction is exothermic.
ii The forward reaction rate will increase because increasing temperature increases the number of collisions with an energy greater than the activation energy.
iii The reverse reaction rate increases for similar reasons to **ii**. (3)

c The high temperature is required to produce an economic rate of production of HCN. The yield must be economic even though a lower temperature would increase the yield. (2)

d The yield and rate must be high enough, so economic, at the stated pressure. Increasing the pressure would be expensive with little rate increase gained. An increase in pressure would lower the yield because there are more product gaseous molecules than reactants. (2)

e **i** +1.

ii Dicyano gold(I) ion.

iii Covalent and dative covalent bonding.

iv $[Au(CN)_2]^-(aq) + Zn(s) \rightarrow Au(s) + Zn^{2+}(aq) + 2CN^-(aq)$. (4)

f When HCl(aq) and KOH(aq) react, only the H^+ and OH^- react to form water because both compounds are fully ionised in water. As HCN is a weak acid it starts partially ionised. All the HCN reacts with the KOH so the unionised HCN ionises, which is an endothermic change, so the overall enthalpy is lower for the HCN with KOH. (2)

g **i**

```
              COOCH3
             /
[—CH2=C—]n
             \
              CN
```

(2)

ii High temperature and high pressure, with an initiator. (2)

(Total 20 marks)

UNIT 28

RECALL TEST

1 See Fig. 28.4. (2)

2 Add aqueous NaOH and the gas coming off should turn pink litmus blue. (3)

3 **a** Br^- ions.

b Possibly CO_2.

c A sulphite (a sulphate(IV) compound).

d A sulphate (sulphate(VI) compound).

e A nitrate. (5)

(Total 10 marks)

CONCEPT TEST

1 **a** The white solid must be dissolved in the minimum amount of hot solvent, filtered hot using vacuum filtration. The fitrate is cooled slowly, then filtered cold. The solid is the purified compound. (4)

b Seal one end of a piece of capillary tube. Tap some of the purified solid into the tube. Attach the tube upright to a thermometer so that the bottom ends are together. Put into an oil bath. Warm it gently, while stirring, and note the temperature at which the crystals melt. Repeat until the readings are the same. (3)

c Put some of the liquid in a flask. Clamp a thermometer so that the bulb is just above the liquid surface. Warm the liquid slowly until it is boiling. Repeat. (2)

2 **a** X: $BaNO_3$; Y: $BaSO_4$; Z: NO_2. (6)

b P: K_2SO_3; Q: SO_2. (2)

c Add aqueous silver nitrate, and nitric acid. A cream (off-white) precipitate would form that would dissolve in concentrated ammonia solution. (3)

(Total 20 marks)

UNIT 29

RECALL TEST

1 **a** mole = mass/RAM.

b mass = mole × RAM.

c RAM = mass/mole. (3)

2 **a** concentration = mole/volume (in dm^3).

b moles = concentration × volume (in dm^3).

c volume (in dm^3) = moles/concentration. (3)

3 **a** **i** $72\,dm^3$.

ii $12\,dm^3$. (2)

b 41.7 moles. (1)

4 Percentage yield = $\dfrac{\text{ACTUAL moles}}{\text{POSSIBLE moles}} \times 100\%$. (1)

(Total 10 marks)

CONCEPT TEST

1 $0.256\,mol\,dm^{-3}$. (2)

2 $I_2(\text{in } KI(aq)) + 2S_2O_3^{2-}(aq) \rightarrow 2I^-(aq) + S_4O_6^{2-}(aq)$.
Concentration iodine = $0.005\,38\,mol\,dm^{-3}$. (3)

3 Concentration chloride ions = $0.0094\,mol\,dm^{-3}$. (3)

4 94.0% pure. (3)

5 $0.0321\,mol\,dm^{-3}$. (3)

6 **a** Percentage yield = 57%. (1)

b Dinitrobenzene was made, probably due to the temperature rising above the 60 °C required for nitration. (2)

7 **a** Conc. $H_2C_2O_4(aq)$ acid = $0.0408\,mol\,dm^{-3}$.

b 0.000 251 moles.

c 0.001 25 moles V^{3+} ions.

d Ratio = 0.201 or 1/4.97.

e Ratio rounds to 1:5 so $MnO_4^-(aq) + 5V^{3+}(aq) + 8H^+(aq) \rightarrow Mn^{2+}(aq) + 5V^{4+}(aq) + 4H_2O(l)$. (8)

(Total 25 marks)

INDEX

Key: Molar mass g mol^{-1} / **Symbol** / Name / Atomic number

Period	Group 1	Group 2											Group 3	Group 4	Group 5	Group 6	Group 7	Group 8
1	1 **H** Hydrogen 1																	4 **He** Helium 2
2	7 **Li** Lithium 3	9 **Be** Beryllium 4											11 **B** Boron 5	12 **C** Carbon 6	14 **N** Nitrogen 7	16 **O** Oxygen 8	19 **F** Fluorine 9	20 **Ne** Neon 10
3	23 **Na** Sodium 11	24 **Mg** Magnesium 12											27 **Al** Aluminium 13	28 **Si** Silicon 14	31 **P** Phosphorus 15	32 **S** Sulphur 16	35.5 **Cl** Chlorine 17	40 **Ar** Argon 18
4	39 **K** Potassium 19	40 **Ca** Calcium 20	45 **Sc** Scandium 21	48 **Ti** Titanium 22	51 **V** Vanadium 23	52 **Cr** Chromium 24	55 **Mn** Manganese 25	56 **Fe** Iron 26	59 **Co** Cobalt 27	59 **Ni** Nickel 28	63.5 **Cu** Copper 29	65.4 **Zn** Zinc 30	70 **Ga** Gallium 31	73 **Ge** Germanium 32	75 **As** Arsenic 33	79 **Se** Selenium 34	80 **Br** Bromine 35	84 **Kr** Krypton 86
5	85 **Rb** Rubidium 37	88 **Sr** Strontium 38	89 **Y** Yttrium 39	91 **Zr** Zirconium 40	93 **Nb** Niobium 41	96 **Mo** Molybdenum 42	99 **Tc** Technetium 43	101 **Ru** Ruthenium 44	103 **Rh** Rhodium 45	106 **Pd** Palladium 46	108 **Ag** Silver 47	112 **Cd** Cadmium 48	115 **In** Indium 49	119 **Sn** Tin 50	122 **Sb** Antimony 51	128 **Te** Tellurium 52	127 **I** Iodine 53	131 **Xe** Xenon 54
6	133 **Cs** Caesium 55	137 **Ba** Barium 56	139 **La** Lanthanum 57	178 **Hf** Hafnium 72	181 **Ta** Tantalum 73	184 **W** Tungsten 74	186 **Re** Rhenium 75	190 **Os** Osmium 76	192 **Ir** Iridium 77	195 **Pt** Platinum 78	197 **Au** Gold 79	201 **Hg** Mercury 80	204 **Tl** Thallium 81	207 **Pb** Lead 82	209 **Bi** Bismuth 83	210 **Po** Polonium 84	210 **At** Astatine 85	222 **Rn** Radon 86
7	223 **Fr** Francium 87	226 **Ra** Radium 88	227 **Ac** Actinium 89															

140 **Ce** Cerium 58	141 **Pr** Praseodymium 59	144 **Nd** Neodymium 60	(147) **Pm** Promethium 61	150 **Sm** Samarium 62	152 **Eu** Europium 63	157 **Gd** Gadolinium 64	159 **Tb** Terbium 65	163 **Dy** Dysprosium 66	165 **Ho** Holmium 67	167 **Er** Erbium 68	169 **Tm** Thulium 69	173 **Yb** Ytterbium 70	175 **Lu** Lutetium 71
232 **Th** Thorium 90	(231) **Pa** Protactinium 91	238 **U** Uranium 92	(237) **Np** Neptunium 93	(242) **Pu** Plutonium 94	(243) **Am** Americium 95	(247) **Cm** Curium 96	(245) **Bk** Berkelium 97	(251) **Cf** Californium 98	(254) **Es** Einsteinium 99	(253) **Fm** Fermium 100	(256) **Md** Mendelevium 101	(254) **No** Nobelium 102	(257) **Lr** Lawrencium 103